EARTH
and PERFORMANCE of
SOLID WASTE LANDFILLS

Proceedings of the session sponsored by the
Soil Mechanics Committee of the
Geotechnical Engineering Division of the
American Society of Civil Engineers
in conjunction with the ASCE Convention in
San Diego, California, October 23-27, 1995

Edited by Mishac K. Yegian and W.D. Liam Finn

Geotechnical Special Publication No. 54

Published by the
American Society of Civil Engineers
345 East 47th Street
New York, New York 10017-2398

ABSTRACT

This proceedings, Construction Congress, consists of papers presented at the 1995 Conference held in San Diego, California from October 22-26, 1995. State-of-the practice and state-of-the-art ideas and presentations are assembled in this volume. It covers a wide range of topics such as ground-water control, outfall construction, design-build delivery systems, productivity, automation, and construction quality assurance.

Library of Congress Cataloging-in-Publication Data

Construction congress : proceedings of the 1995 conference, San Diego, California, October 2226, 1995 / sponsored by the Construction Division of the American Society of Civil Engineers ; edited by C. William Ibbs.
    p.   cm.
    ISBN 0-7844-0128-4
    1. Civil engineering—Congresses. 2. Construction industry—Congresses. 3. Building-Congresses.I. Ibbs, C. William. II. American Society of Civil Engineers. Construction Division.
TA5.C926   1995   95-41125
624—dc20   CIP

GEOTECHNICAL SPECIAL PUBLICATIONS

1) TERZAGHI LECTURES
2) GEOTECHNICAL ASPECTS OF STIFF AND HARD CLAYS
3) LANDSLIDE DAMS: PROCESSES RISK, AND MITIGATION
4) TIEBACKS FOR BULKHEADS
5) SETTLEMENT OF SHALLOW FOUNDATION ON COHESIONLESS
   SOILS: DESIGN AND PERFORMANCE
6) USE OF IN SITU TESTS IN GEOTECHNICAL ENGINEERING
7) TIMBER BULKHEADS
8) FOUNDATIONS FOR TRANSMISSION LINE TOWERS
9) FOUNDATIONS AND EXCAVATIONS IN DECOMPOSED ROCK OF
   THE PIEDMONT PROVINCE
10) ENGINEERING ASPECTS OF SOIL EROSION DISPERSIVE CLAYS AND LOESS
11) DYNAMIC RESPONSE OF PILE FOUNDATIONS— EXPERIMENT,
    ANALYSIS AND OBSERVATION
12) SOIL IMPROVEMENT - A TEN YEAR UPDATE
13) GEOTECHNICAL PRACTICE FOR SOLID WASTE DISPOSAL '87
14) GEOTECHNICAL ASPECTS OF KARST TERRIANS
15) MEASURED PERFORMANCE SHALLOW FOUNDATIONS
16) SPECIAL TOPICS IN FOUNDATIONS
17) SOIL PROPERTIES EVALUATION FROM CENTRIFUGAL MODELS
18) GEOSYNTHETICS FOR SOIL IMPROVEMENT
19) MINE INDUCED SUBSIDENCE: EFFECTS ON ENGINEERED STRUCTURES
20) EARTHQUAKE ENGINEERING & SOIL DYNAMICS (II)
21) HYDRAULIC FILL STRUCTURES
22) FOUNDATION ENGINEERING
23) PREDICTED AND OBSERVED AXIAL BEHAVIOR OF PILES
24) RESILIENT MODULI OF SOILS: LABORATORY CONDITIONS
25) DESIGN AND PERFORMANCE OF EARTH RETAINING STRUCTURES
26) WASTE CONTAINMENT SYSTEMS; CONSTRUCTION, REGULATION, AND
    PERFORMANCE
27) GEOTECHNICAL ENGINEERING CONGRESS
28) DETECTION OF AND CONSTRUCTION AT THE SOIL/ROCK INTERFACE
29) RECENT ADVANCES IN INSTRUMENTATION, DATA ACQUISITION AND
    TESTING IN SOIL DYNAMICS
30) GROUTING, SOIL IMPROVEMENT AND GEOSYNTHETICS
31) STABILITY AND PERFORMANCE OF SLOPES AND EMBANKMENTS II
    (A 25 YEAR PERSPECTIVE)
32) EMBANKMENT DAMS-JAMES L. SHERARD CONTRIBUTIONS
33) EXCAVATION AND SUPPORT FOR THE URBAN INFRASTRUCTURE
34) PILES UNDER DYNAMIC LOADS
35) GEOTECHNICAL PRACTICE IN DAM REHABILITATION
36) FLY ASH FOR SOIL IMPROVEMENT
37) ADVANCES IN SITE CHARACTERIZATION: DATA ACQUISITION,
    DATA MANAGEMENT AND DATA INTERPRETATION
38) DESIGN AND PERFORMANCE OF DEEP FOUNDATIONS: PILES AND
    PIERS IN SOIL AND SOFT ROCK
39) UNSATURATED SOILS
40) VERTICAL AND HORIZONTAL DEFORMATIONS OF FOUNDATIONS AND
    EMBANKMENTS
41) PREDICTED AND MEASURED BEHAVIOR OF FIVE SPREAD FOOTINGS ON SAND
42) SERVICEABILITY OF EARTH RETAINING STRUCTURES
43) FRACTURE MECHANICS APPLIED TO GEOTECHNICAL ENGINEERING
44) GROUND FAILURES UNDER SEISMIC CONDITIONS
45) IN-SITU DEEP SOIL IMPROVEMENT
46) GEOENVIRONMENT 2000
47) GEO-ENVIRONMENTAL ISSUES FACING THE AMERICAS
48) SOIL SUCTION APPLICATIONS IN GEOTECHNICAL ENGINEERING
49) SOIL IMPROVEMENT FOR EARTHQUAKE HAZARD MITIGATION
50) FOUNDATION UPGRADING AND REPAIR FOR INFRASTRUCTURE IMPROVEMENT
51) PERFORMANCE OF DEEP FOUNDATIONS UNDER SEISMIC LOADING
52) LANDSLIDES UNDER STATIC AND DYNAMIC CONDITIONS - ANALYSIS,
    MONITORING, AND MITIGATION
53) LANDFILL CLOSURES - ENVIRONMENTAL PROTECTION AND LAND RECOVERY
54) EARTHQUAKE DESIGN AND PERFORMANCE OF SOLID WASTE LANDFILLS
55) EARTHQUAKE - INDUCED MOVEMENTS AND SEISMIC REMEDIATION OF
    EXISTING FOUNDATIONS AND ABUTMENTS
56) STATIC AND DYNAMIC PROPERTIES OF GRAVELLY SOILS

PREFACE

Seismic design considerations for municipal solid waste landfills continue to receive increased attention in geotechnical engineering practice. The US Code of Federal Regulation (USEPA Subtitle D) requires that all landfills within a "seismic impact zone" be designed for a level of acceleration of not exceeding 90% or greater probability in 250 years (return period of 2373 years). A "seismic impact zone" is defined as a region within which the peak horizontal acceleration for the stated probability level is greater than 0.1 g. A seismic hazard map prepared by U.S.G.S. (1991) shows the levels of acceleration to be considered for landfills across the United States. From this map, it can be observed that landfills located in most parts of the Country will have to be designed for earthquakes. The State of California requires that all landfills be designed to withstand the maximum probable or a maximum creditable earthquake.

Clearly, there is urgency in geotechnical engineering practice to address seismic design issues related to landfills. During the past few years, researchers and practitioners have been developing improved understanding of landfill response under earthquake excitations. The Northridge earthquake created invaluable opportunity to further enhance this developing knowledge.

This Special Technical Publication (STP) contains twelve papers that summarize current state-of-practice and knowledge related to seismic design and performance of landfills. The papers cover topics that include: ground motions for landfill design; seismic response analysis of landfills; dynamic response analysis of geosynthetic systems; dynamic properties of landfill wastes, regulatory guidelines; performance of landfills during recent earthquakes; and future research needs.

It is the current practice of the Geotechnical Engineering Division that each paper published in a STP be reviewed for its content and quality. These special technical publications are intended to reinforce the programs presented at convention sessions or specialty conferences and to contain papers that are timely and may be controversial to some extent. Because of the need to have the STP available at the convention, time available for reviews is generally not as long and reviews may not be as comprehensive as those given to papers submitted to the Journal of the Division. Therefore, it should be recognized that there is difference in the purpose and technical status of contributions to the special technical publications as compared to those in the Journal. In accordance with ASCE policy, for these Proceedings, each paper received at least two independent positive peer reviews. All papers published in this volume are eligible for discussion in the Journal of the Geotechnical Engineering Division and are eligible for ASCE awards.

The session was planned and organized by Mishac Yegian, Liam Finn, Shobha Bhatia, Steve Kramer and Panos Dakoulas (Chair) of the Soil Dynamics Committee. Gratitude is expressed to the following individuals who peer-reviewed the abstracts and the manuscripts of the papers, under an accelerated time frame:

Hassan Abedi	Liam Finn	Richard Reynolds
Shobha Bhatia	Vahé Ghahraman	Mishac Yegian
Panos Dakoulas	Steve Kramer	
Pedro DeAlba	Madjid Lahlaf	
Donald Del Nero	Catalina Marlunda	

v

Also, appreciation is expressed to the authors for contributing to this volume and for their expeditious responses to the reviewers' comments.

Finally, the session organizers and the Soil Dynamics Committee of the Geotechnical Engineering Division of ASCE thank Shiela Menaker and the staff of ASCE for their patience and for arranging the printing of this Special Technical Publication.

Mishac K. Yegian
Professor and Chairman
Department of Civil Engineering
Northeastern University
Boston, MA 02115

W. D. Liam Finn
Professor of Civil Engineering
Civil Engineering Department
University of British Columbia
Vancouver, BC V6T 1Z4 Canada

Proceedings Editors

CONTENTS

EARTHQUAKE DESIGN & PERFORMANCE OF SOLID WASTE LANDFILLS

Earthquake Related Damage and Landfill Performance

Robert L. Anderson[1]

Abstract

Staff of the California Integrated Waste Management Board (IWMB) have monitored landfill damage in response to earthquakes after several events beginning in late 1987 and, most recently, after the Northridge earthquake of 1994. Monitoring has consisted of visual inspections of landfill systems in the days following the event. Of the various landfills visited by IWMB staff after earthquakes, no catastrophic slope or foundation failures related to earthquakes have been observed. However, as was to be expected, damage of varying degrees has been noted. Where observed, such damage has typically impacted the cover and/or environmental control systems. In some instances, earthquake damage to landfill systems may not be apparent during post earthquake visual inspections but may appear some time later.

This paper will discuss the types of damage that have been observed and some of the potential impacts to landfill performance that may result. Observation of landfill performance in response to earthquakes has aided IWMB staff in the review of design and closure reports submitted by the regulated community. The presentation of this information may assist owner/operators, consultants, and other regulators in assessing the performance of landfills in response to earthquakes and may also aid in the design of more earthquake-resistant landfill systems.

Introduction

The primary purpose for IWMB staff to visit landfills after earthquakes is to obtain reconnaissance level information as to whether the landfill can continue to operate (often at increased daily tonnages due to an increase in the waste stream after an earthquake) or if the facility must be closed to repair damage caused by earthquake-related effects prior to reopening. Only a limited number of written accounts of landfill performance during earthquakes have been transmitted to the IWMB. As a result, an effort has been made to increase the number of accounts by encouraging operators to prepare written reports of landfill damage and performance; and by sending teams of engineering geologists and engineers to sites

---

[1] Associate Engineering Geologist, California Integrated Waste Management Board, 8800 Cal Center Drive, Sacramento, CA 95826

1

near the epicenters of significant California earthquakes to make observations that may lead to consideration of design, construction, or operational changes that could minimize earthquake damage in the future.

A listing of the landfills discussed and the distance to the epicenter(s) of the earthquake(s) is listed in Table 1. A brief description of the earthquakes mentioned below is presented in Table 2.

TABLE 1. General Parameters

LANDFILL NAME	DISTANCE	MMI AT SITE	SITE GEOLOGY	LANDFILL TYPE	IMPACT

NORTHRIDGE EARTHQUAKE

LANDFILL NAME	DISTANCE	MMI AT SITE	SITE GEOLOGY	LANDFILL TYPE	IMPACT
Calabasas	15 km	VII	HR	PF/AF	Moderate
Sunshine Canyon	14 km	VIII	HR	CF	Moderate
Chiquita Canyon	19 km	VII	SR	AF/CF	Moderate
Toyon Canyon	23 km	VII	HR	CF	Minor
Bradley Ave. East	15 km	VII	SR	PF	Minor
Bradley Ave. West	15 km	VII	HR	CF	Moderate
Lopez Canyon	17 km	VIII	HR	CF	Moderate
Russell Moe	16 km	VIII	HR	CF	Moderate
Tierra Rejada	28 km	VII	HR	SH/CF	Minor
Simi Valley	27 km	VII	HR	AF/CF	None
Scholl Canyon	34 km	VI	HR	CF	Moderate
Puente	54 km	VI	HR	CF	None
Savage Canyon	56 km	VI	HR	SH/CF	None
Palmdale	48 km	VI	SH-HR	SH/CF	Minor
Azusa	60 km	V	SR	PF	None
Spadra	70 km	IV-V	SR-HR	SH/CF	None

TABLE 1. (continued)

LANDFILL NAME	DISTANCE	MMI AT SITE	SITE GEOLOGY	LANDFILL TYPE	IMPACT
SAN FERNANDO EARTHQUAKE					
Sunshine Canyon	13 km	VIII	HR	CF	Minor
Toyon Canyon	28 km	X	HR	CF	None
Bradley Ave, East	12 km	X	SR	PF	None
Scholl Canyon	29 km	VII	HR	CF	None
Calabasas	40 km	VII	HR	PF/AF	None
Russell Moe	08 km	X	HR	CF	Minor
LOMA PRIETA EARTHQUAKE					
Redwood	120 km	V	BM	AF	Minor
Guadalupe	20 km	VII	HR	AF	Minor
Santa Cruz	20 km	VII	SR	CF	Minor
Zanker Road	43 km	VI	SR	AF	Minor
SANTA ROSA EARTHQUAKE					
Redwood	40 km	V	BM	AF	Moderate
LANDERS EARTHQUAKE					
Landers	07 km	VIII	SR	AF	Minor

LEGEND

SITE GEOLOGY
BM - Bay Mud
SR - Soft Rock
HR - Hard Rock
MMI - Modified Mercalli
     Intensity

LANDFILL TYPE
AF - Area Fill
CF - Canyon Fill
PF - Pit Fill
SH - Side Hill

TABLE 2. Earthquakes

DATE	NAME	NATURE OF MOTION	MAGNITUDE	ASSOCIATED FAULT
10/01/69	Santa Rosa	Right Lateral	5.6 and 5.7 M_L	Healdsburg
02/09/71	San Fernando	Thrust and Left Lateral	6.4 M_S	San Fernando
10/17/89	Loma Prieta	Thrust and Right Lateral	6.9 M_W	San Andreas
06/28/92	Landers	Right Lateral	7.3 M_W	Johnson Valley
01/17/94	Northridge	Thrust	6.7 M_W	Oak Ridge

LEGEND

M_S = Surface Magnitude　　　M_W = Moment Magnitude　　　M_L = Local Magnitude

For the purposes of this report, municipal solid waste landfills are categorized as area fills, pit fills, cut and cover side-hill canyon fills, and canyon fills. Area fills may or may not have a liner and leachate collection and removal system. In constructing area fills, excavation is limited to that necessary to install features such as liners. This type of fill operation is one of the most common in California. Pit fills were once a popular method to reclaim sand and gravel quarries in southern California. Fill is placed in a pit until a set elevation is reached, usually slightly above the pre-excavation grade elevation. Cut and cover side hill canyon fills are landfills where part of a canyon is used for refuse disposal, while either a ridge or a borrow pit is used for a source for cover. The canyon fill is a landfill where a canyon is partially or completely filled vertically, while being filled across the breadth of the canyon. During operations at area fills, the crown and the toe of fill slopes may crack and differentially settle due to either consolidation or strong ground shaking. Pit fills may have dynamic compaction occur along the edge of the fill. Both cut and cover side hill and canyon fills may have dynamic compaction and/or differential settlement occur due to strong ground shaking related to earthquakes. Area fills differ from the other types of municipal solid waste landfills mentioned mainly by the lack of lateral support for the refuse mass and cover provided in various degrees by either canyon walls or excavations. Shear wave propagation through area fills differs from pit and canyon fills since the medium for transmission of the wave is through the base of the landfill for area fills and may be both through the base and (at least) part of the canyon wall or excavation adjacent to the side of the canyon or pit fill, respectively.

The Northridge earthquake occurred on January 17, 1994. Between January 19 and 27, 1994, the IWMB sent two teams to observe performance and/or damage to municipal solid waste landfills in the general vicinity of the epicenter of the Northridge earthquake. Sites visited were selected based upon:

1. The proximity of the site to the epicenter;
2. Design and operating practices at the site; and
3. IWMB staff familiarity with the sites.

The following summaries are based on observations made by IWMB staff during site visits to nine of fifteen facilities visited and a review of landfill performance assessments submitted to the IWMB by landfill operators. The nine sites were selected based upon the type and amount of damage observed by IWMB staff that was attributed to the earthquake. This was the third earthquake after which the IWMB sent a field team made of engineers and/or engineering geologists to observe landfill performance after an earthquake. Because several of the landfills visited were near the epicenter of the 1971 San Fernando earthquake, a comparison of the performance during the two events is included for some sites.

Calabasas Landfill, Los Angeles County

The 416 acre Calabasas Landfill was the closest (15 km) active landfill to the epicenter of the Northridge earthquake. The partially lined site is underlain by landslide deposits, interbedded sandstone and shale of the Topanga Formation, and interbedded sandstone and conglomerate of the Modelo Formation. Both the Topanga and Modelo Formations are Miocene in age. Three faults have been identified on site. The faulting is apparently associated with the development of the folding which may have ceased locally in the Miocene. None of the faults are known to have been active in the past 11,000 years (Holocene). Augello and others (1995) have estimated that the site rock peak horizontal acceleration associated with the main shock of the Northridge earthquake was 0.20g.

A cracked concrete drainage ditch was reported by the site superintendent. A 215-meter long tension crack was observed along the northern edge of the northwestern fill area. The tension crack was sub-parallel to the crown of the anchor trench on the fill side of a lined cell and was locally up to 15 cm wide and vertically offset up to 10 cm down on the fill side. No refuse or liner materials were exposed. Several additional cracks were observed on both the roadway benches and fill slopes on the west side of the site. The cracks had relatively sharp edges and ran both parallel and perpendicular to the benches. The perpendicular cracks ran partially down the slopes. No bulging of fill was observed either parallel or perpendicular below the cracks.

Sunshine Canyon Landfill, Los Angeles County

The unlined, 148 acre Sunshine Canyon landfill has not received waste since September 1991 and is awaiting closure. The site is located approximately 13.6 km and 13 km from the epicenters of the 1994 Northridge and the 1971 San Fernando earthquakes, respectively. Augello and others (1995) have estimated that the site rock peak horizontal acceleration associated with the main shock of the Northridge earthquake was 0.46g. The side slopes range from 2:1 to 1.75:1 (horizontal to vertical). Well-indurated sandstones with siltstones, mudstones, and conglomerates of the Miocene to lower Pliocene age Towsley Formation make up the bedrock beneath the site. Several pre-Holocene fault splays are located within the site boundaries. Several landslides occurred in the vicinity of the site as a result of strong ground shaking from both the Northridge and San Fernando earthquakes. IWMB records indicate that the site had a shallow slump in a fill slope as shown in a aerial photograph taken shortly after the San Fernando earthquake.

Tension cracking was observed parallel to the crown of the slope. Localized tension cracks were observed extending up to 4.6 meters back from the crown of the slope. The eastern end of the fill slope was cracked.

A settlement crack was observed in the top deck running sub-parallel to the southern cut slope. The crack was vertically offset up to 38 cm down on the north (fill) side and was approximately 100 meters in length. Refuse was not exposed in the settlement crack. Minor damage was reported by the operator to the bottom of a small, above-ground water tank located on native ground. The cracked fill and the damage to the water tank are attributed to strong ground shaking.

Chiquita Canyon Landfill, Los Angeles County

The partially lined 154 acre Chiquita Canyon landfill is underlain by alluvium; soft siltstone interbedded with mudstone of the Pliocene age Pico Formation; poorly to moderately indurated sandstone, siltstone, mudstone and conglomerate of the Pliocene to Pleistocene age Saugus Formation; and recent landslide deposits. No active faults are known to cross the landfill. Augello and others (1995) have estimated that the site rock peak horizontal acceleration associated with the main shock of the Northridge earthquake was 0.33g.

The facility is divided into five waste management units, known as areas or canyons "A","B","C","D", and the primary canyon. Waste management areas "A" and "D" are separated by a soil fill wedge. Cracking near the contact between area "D" and the soil fill had localized displacement of up to 20 cm vertically and 30 cm horizontally. Cracking of soil was observed up to 12 cm vertically and 15 cm horizontally in cover on the waste side of the contact between the south side of waste management area "A" and the soil fill prism.

A tear in the 60-mil High Density Polyethylene liner element was observed near the top of the western edge of waste management area "C." The tear was 4.3 meters long and up to 25 cm wide and ran parallel to the hinge point of the anchor trench. The clay beneath the torn geomembrane did not appear to have been cracked.

Area "B" had some minor settlement along the perimeter of the fill-native ground contact but no waste was exposed. The earthen roadway adjacent to the site's flare station was cracked, but the flare station itself appeared to be undamaged. The flare station was off-line for about 10 hours due to a loss of power.

Toyon Canyon Landfill, Los Angeles County

Toyon Canyon Landfill is a 90-acre, unlined canyon fill located in the Santa Monica Mountains. The site is underlain by Miocene age sandstone and conglomerate of the Hollycrest Formation. The Griffth Fault bounds the site on the southern side. No known active faults cross the site. Augello and others (1995) have estimated that the site rock peak horizontal acceleration associated with the main shock of the Northridge earthquake was 0.21g. No refuse has been accepted at the site since February 1986. Fill slopes at the site vary from 2:1 to 3:1 (horizontal to vertical). Total fill relief at the site is 430 feet. Cracking and settlement in fill along the contact line between fill and native materials was observed along the edges of several of the benches. A compression ridge found on the western slope of the ridge was 2 cm high and 21.5 meters long. The header to one landfill gas collection well had been sheared by strong ground shaking. The site's gas-to-energy facility was off-line for approximately 27 hours after the earthquake due to a loss of electrical power.

The operator of the landfill gas-to-energy facility prepared a letter report on the damage to the gas extraction system. The operator had the landfill gas extraction wells at the site sounded and determined that 48 wells had a decrease in well depth after the earthquake. The wells were last sounded more than a year before the earthquake. In some of the wells, the decrease may have occurred prior to the earthquake from non-earthquake-related causes, but at least some are due to earthquake-related effects. Landfill gas composition and flow rates were compared before and after the Northridge earthquake. The operator indicates that there was less than a 5% change in flow rate and in the composition of the gas. Settlement cracking along the perimeter of the canyon may have been caused by dynamic compaction of the refuse due to strong ground shaking. The high number of damaged landfill gas extraction wells may be due in part to shearing or crushing during the earthquake.

Bradley Avenue Landfills East, West, and West Extension, Los Angeles County

The 209 acre Bradley Avenue Landfills East, West, and West Extension facility is located in an old gravel quarry approximately 15 km east of the epicenter of the Northridge earthquake. Bradley Avenue East was in operation during the San Fernando earthquake but no records of damage or performance are available in IWMB files. The Bradley Avenue East Landfill is unlined and was partially closed at the time of the Northridge earthquake. Site offices and a sortline for recycling are located on the landfill. No damage to the sortline or to the site office was reported by the operator. IWMB staff were on site in the operator's offices during four Richter magnitude 4+ aftershocks on January 21, 1994. The aftershocks did not create any additional observable surface damage. Anderson and Kavazanjian (1995) estimated that the free field peak horizontal ground acceleration at the Bradley Avenue Landfills was 0.45g.

Settlement cracks were observed along the western side of the flare station pad at the Bradley Avenue East Landfill. The settlement of the fill adjacent to the flare station pad and nearby gas collection lines did not affect the performance of the flare station once power was restored to the site. The facility's landfill gas flare station is located along the eastern edge. Soil had settled up to 15 cm along the western side of the flare station.

Bradley Avenue West Extension Landfill had extensive cracking and localized settlement of fill along the western edge of the pit. A tear in the geotextile component of the liner was observed near the end of the fill. The tear had supposedly begun before the earthquake but was enlarged by the earthquake. No damage to the underlying geomembrane was observed. Damage to the surficial fill over the liner and the anchor trenches may be due to the lower degree of compaction compared to the in-place native alluvial gravels and sands.

Lopez Canyon Landfill, Los Angeles County

This partially lined 166 acre canyon fill facility is located along the eastern edge of the mouth of Lopez Canyon. Solid waste disposal operations began at the site in 1975. Geologic units beneath the site include siltstones, sandstones, and conglomerates of the Modelo, Pico, Towsley, and Saugus Formations. The Oak Hill, Kagel and Tujunga Faults are located to the northwest and through the southeast corner of the site, respectively. During the San Fernando earthquake, surface rupture due to fault movement occurred on the Oak Hill, Kagel, and the

Tujunga faults near, but not through the site. After that earthquake, several minor slides were noted in native soil near the site. In addition, soil on the top several native ridge tops had been shattered and cracked. However, no damage was noted at the landfill since the site did not exist. Anderson and Kavazanjian (1995) estimated that the free field peak horizontal ground acceleration at the Lopez Canyon Landfill associated with the Northridge earthquake was 0.44g.

On January 20, 1994, IWMB staff toured areas "A", "B", "AB+", and "C." An engineer from the City of Los Angeles Bureau of Sanitation indicated that the pipe works attached to the bottom of a one-million gallon water tank near the flare station had broken. The water in the tank flowed out and down slope towards the flare station and the south slope of area "B." No significant erosion occurred on the slope and the water was diverted and contained by the sediment basins on-site. Cracking was observed along the top of the fill slope of area "B." Shallow slumping had occurred on the slope prior to the earthquake. The flare station was off-line for a period of 13 hours as a result of the loss of power to the site. Several landfill gas collection lines had broken on the north side of area "B." A tear in a 80-mil high density polyethylene liner in area "C" had been reported by an engineer working at the site. The slope on which the liner tear was indicated to have occurred on was approximately 1:1 (horizontal to vertical). The tear was reported to have been repaired prior to IWMB staff arrival. The supposed location of the tear lead IWMB staff to believe that the tear was caused by strong ground shaking and not by installation or operating practices as had been suggested by the operator.

Russell Moe Landfill, Los Angeles County

This 15 acre unlined landfill is located on the edge of the western ridge at the mouth of Lopez Canyon. Sandstone, shale, and conglomerate of the Modelo and Pico Formations underlie the site. The Oak Ridge, Tujunga, and Kagel faults border the site to the north, south, and east. The site was closed in 1961 and a trailer park was constructed over the eastern edge of the fill. Several one story warehouses and office buildings are located near the west central section of the landfill. During the 1971 San Fernando earthquake, some differential settlement occurred at the site along the limits of the fill. The estimated peak horizontal ground acceleration during the Northridge earthquake was 0.44g based upon the closest California Strong Motion Instrument Program (CSMIP) seismic monitoring station, located in Kagel Canyon, approximately 3 km from the site.

Two gas mains were ruptured during the earthquake and the trailer court was temporarily evacuated. Trailers located on and off the landfill were knocked off their foundations. The occupants of the buildings located near the middle of the landfill indicated that no significant damage occurred during the earthquake. Cracking occurred along the projected eastern edge of the landfill and native ground. Locally, native ground and roadways were ruptured. Damage to the trailer park was of similar magnitude both off and on the landfill.

Tierra Rejada Landfill, Ventura County

This 25 acre unlined canyon landfill was closed in 1972. The site was visited after the Northridge earthquake because there had been a historic slope instability problem and because of its proximity (28 km north) to the epicenter of the Northridge earthquake. The site is located along the southern bank of Arroyo Simi and is underlain by massive sandstone and interbedded claystones of the Oligocene age Sespe Formation. Two minor faults have been previously mapped at the site. Augello and others (1995) have estimated that the site rock peak horizontal acceleration associated with the main shock of the Northridge earthquake was 0.21g. Five slope inclinometers were installed at the site as part of a stability evaluation and monitoring program prior to the Northridge earthquake. As part of a corrective action program to stabilize a fill slope, an armored buttress fill was placed on the northern side of the site adjacent to Arroyo Simi, and the Arroyo Simi was diverted around the toe of the fill.

In the southwestern portion of the landfill, a series of cracks ranging from 1 to 15 cm in width were discovered. Several 2 to 7 cm wide cracks were also observed along the margins of the fill areas. Landfill gas ranging from 500 to 1,500 ppmv was detected in cracks on the western side of the site. The western cracks did not have crisp edges as did cracks in the benches, the southwestern area, or the top deck of the site, indicating that some of the cracking observed occurred before the Northridge earthquake. Cracking in the top deck appeared to be in alignment with the minor faults and with a small rock fall near the southern edge of the fill area. The northern side of the cracks were offset downward by 15 cm. The site's inclinometers were checked. The inclinometer readings indicated that the refuse had not moved appreciably. It is not clear if the cracking on the top deck was caused by dynamic compaction or by fill rupture caused by fault displacement through the site. No damage to the armored fill buttress was noted by IWMB staff.

The following discussion is based upon IWMB file research on landfill performance information on the Redwood landfill during the Santa Cruz and Loma Prieta earthquakes and the Landers Solid Waste Disposal Facility during the Landers earthquake.

Redwood Landfill, Marin County

The 200-acre Redwood Landfill is located upon a shallow deposit of young Bay Muds and alluvium overlying Franciscan Bedrock. The site is partially lined with clay. Notes regarding landfill damage caused by the Santa Rosa earthquakes of October 1, 1969, and IWMB file notes of the landfill performance during the Loma Prieta earthquake of October 17, 1989, have been reviewed by IWMB staff. O'Rourke (1973) reported that, during the Santa Rosa earthquakes, a few interior cell walls made up of clay (Bay Muds) collapsed, but the perimeter levee around the landfill was not damaged. Orr and Finch (1990) reported minor damage to a landfill gas line pipe joint as a result of strong ground shaking from the Loma Prieta earthquake.

Static failures occurred in the perimeter levee in December 1988 and in 1992. Reports of landfill performance from both the Santa Rosa and the Loma Prieta earthquakes do not indicate that the existing perimeter levees were damaged.

Modified Mercalli Intensities (MMI) for areas around the site were the same for both earthquakes, even though the magnitude of the earthquakes were not similar. Estimated peak horizontal ground accelerations at the site from the October 1969, earthquakes are 0.05g. Estimated peak horizontal ground acceleration at the site from the October 1989 earthquake is 0.08g. Both peak horizontal ground acceleration estimates are based upon horizontal acceleration versus distance curves found in Krinitzsky and others (1988). Similarity in MMI values may be partially explained by the lower magnitude earthquake epicenter being closer than the larger earthquake. The site may have had more damage from the earlier earthquake due to the short time between the two similar sized main shocks, the use of Bay Muds to build the interior cell walls, and the wavelength to period ratio of the shear wave to the site.

Landers Landfill, San Bernardino County

Landers Solid Waste Disposal Facility is a 44-acre unlined landfill located approximately 15 km north of the town of Yucca Valley and 5 km east from the epicenter of the Landers earthquake. Geology at the site consists of shallow wind blown sands and alluvium overlying fractured quartz monzonite and gneiss. Two pre-Holocene faults run beneath the site.

The site's setting is unique in that the Landers earthquake resulted in extensive surface ground rupture 5 km west and 8 km northwest of the site, along the Johnson Valley and Homestead Valley faults, respectively.

The estimated horizontal ground acceleration at the site was 0.26g. The closest California Strong Motion Instrumentation seismic station was located in Joshua Tree, 8 km southeast of the site (14 km southeast of the epicenter). The peak horizontal ground acceleration measured at the Joshua Tree station was 0.29g (DMG OSMS Report 92-09).

A special site inspection was conducted on June 29, 1992. During the inspection, cracks were observed in the fill side slopes of the refuse cell closest to the septage ponds. This is an area where dried sludge had been placed over refuse, and then covered with uncompacted soil. A slope failure was also observed on-site but was attributed to erosion of the slope by water.

Discussion

Damage to landfills observed by IWMB staff is categorized into four groups: 1. cracking of daily, intermediate, or final covers; 2. damage to liners; 3. damage to environmental collection and control systems; and 4. damage to infrastructure such as water tanks and on-site structures. Each of the categories is discussed below:

Cracking observed at sites can be divided into lateral tension, settlement, and compression. The kind of displacement manifested as tension cracking is not unlike the settlement observed in unconsolidated earthen fill in drainage projects as reported by California Division of Mines and Geology (DMG) staff in Simi Valley as a result of the Northridge earthquake (DMG Open File Report 94-09). The most common type of cracking observed by the IWMB was lateral tension cracks. Lateral tension cracks are tension cracks with no appreciable vertical off-set. These cracks were observed near the crown of fill slopes.

Settlement cracks were observed mainly along the edges of fills such as the fill and cut or native soil contact, when off-set was down on the fill side. Settlement cracks were observed above the inner edge of anchor trenches and in the top deck of Sunshine Canyon. Only two compression cracks were observed, one on the west fill slope at Toyon Canyon Landfill and the other in a fill area north of the main waste management unit at Sunshine Canyon. Compression cracks bulged upwards 2 cm and were over 10 meters in length. Fill rupture (soil that is both laterally and vertically cracked and offset and not parallel to a slope, edge of fill or anchor trench) was observed in intermediate soil cover over relatively shallow refuse at the Chiquita Canyon landfill.

The most significant amount of damage observed by IWMB staff after the Loma Prieta earthquake was to the 39.6 acre Santa Cruz Landfill. Johnson and others (1991) estimated that the peak horizontal ground acceleration at the Santa Cruz Landfill associated with the main shock of the Loma Prieta earthquake was 0.45g. Damage to the Santa Cruz landfill was confined to cracking of the fill slopes and benches parallel to the crown of the slopes. Cracking was also noted in a dike around the septage and leachate ponds. The amount of damage at the Santa Cruz Landfill was not as extensive as the amount of damage observed at the Calabasas,

Sunshine, Chiquita, and Bradley Avenue West Extension landfills. This may be due to three factors:

1.   Type of faulting associated with the earthquakes;
2.   The geology near and at the sites; and
3.   The locations of the landfill related to the attitude of the area rupture associated with the earthquake under study.

Other than tearing of the synthetic liner at Chiquita Canyon Landfill and tearing of the geotextile element of the Bradley Avenue West liner, distress of exposed liner materials was not observed by IWMB staff after the Northridge earthquake. Three factors contributed to limiting observations of liner distress. First, many of the landfills within areas of high ground shaking are either unlined or only partially lined. Second, some of the liners are covered and were not observed or easily monitored by operators. Third, some preexisting damage to geotextile materials in liners was not well documented prior to the earthquake.

Landfill gas collection lines, headers, and flare stations were out of service for one of two reasons: 1. the power was out to the flare stations; or 2. the headers, gas collection lines, or gas extraction wells had been damaged. The landfill gas collection systems were operable after power was restored to the sites, and the systems were checked by the operators. The overall landfill gas collections and control systems were up and operating although several landfills had leaking pipe joints, broken lateral landfill gas collection lines, broken headers, or gas extraction wells with decreased well depths.

Secondary structures, such as water tanks and trailers, were affected by the Northridge earthquake, and required more repairs than did most slopes, with the exception of Chiquita Canyon Landfill. Strong ground shaking cracked the bottoms of water tanks at the Toyon Canyon Landfill, and Sunshine Canyon Landfill, and ruptured the pipe works attached to the bottom of the water tank at the Lopez Canyon Landfill.

It should be noted that trailers located on the Zanker Road and the Guadalupe Landfills were knocked off their foundations (Buranek and Prasad 1991) during the Loma Prieta earthquake. The estimated peak horizontal ground accelerations at the Zanker Road Landfill and the Guadalupe Landfill was 0.23g and 0.35g, respectively (Buranek et al.). Additionally, some trailers at the Lopez Canyon Landfill and the Russell Moe Landfill were knocked off of their foundations during the Northridge earthquake. Dislocation of trailers in the case of both the Loma Prieta and the Northridge earthquakes was not seen as a unique feature to landfills since some trailers located off-site of the landfills were knocked from their foundations.

Damage to landfills resulting from the Loma Prieta, Landers, and Northridge earthquake aftershocks has not been reported to the IWMB.

Conclusions

Daily, intermediate, and final cover over municipal solid waste was locally cracked and displaced in a similar manner as some of the channel fills reported by the DMG in the Simi Valley area. Some of cracking may have been caused by dynamic compaction. Of the damage observed, the most significant was associated, not with the surficial fill, but with landfill gas collection lines and water tanks. Information regarding damage and landfill performance after the Loma Prieta, Landers, and the Northridge earthquakes has made it clear that a systematic assessment of landfills and related infrastructure should continue to be developed and followed.

Acknowledgements

I would like to thank Ms. Charlene Herbst, Mr. John Clinkenbeard, Mr. Darryl Petker, Ms. Duv Holland, Ms. Donna Khalili, Ms. Vicki Hanson, and Ms. Maria Carrillo of the Integrated Waste Management Board for their encouragement, review, and support both in the field and in the office. I would also like to thank Dr. Jonathan Bray of the University of California at Berkeley for his encouragement and advice.

References

Anderson, D. G., and Kavazanijan E., (1995). "Performance of Landfills Under Seismic Loading" in State of the Art Report, International Conference on Case Histories in Geotechnical Engineering, Rolla Missouri.

Augello, A.J., Bray, J.D., Matasovic, N., Kavazanjian, E., Seed, R.B., (1995). "Solid Waste Landfill Performance During the 1994 Northridge Earthquake" in State of the Art Report, International Conference on Case Histories in Geotechnical Engineering, Rolla Missouri.

Barrows, A. G., Siang, S.T., Irvine, P.J. (1994). "Investigation of Surface Geologic Effects and Related Land Movement in the City of Simi Valley Resulting From the Northridge Earthquake of January 17, 1994." California Division of Mines and Geology Open File Report 94-09.

References

Buranek, D., and Prasad, S. "Sanitary Landfill Performance During the Loma Prieta Earthquake," in Proceedings: Second International Conference on Recent Advances in Geotechnical Earthquake Engineering and Soil Dynamics, March 11-15, 1991, St. Louis Missouri.

CDMG (1990). "The Loma Prieta (Santa Cruz Mountains), California Earthquake of 17 October 1989, "California Division of Mines and Geology Special Publication No. 104", pages 14 and 15.

CIWMB (1994), "Observations of Landfill Performance, Northridge Earthquake of January 17, 1994," 4 p. (plus attachments).

Colorado River Basin Regional Water Quality Control Board, Landers Class III Landfill, Board Order No. 91-028, letter to San Bernardino County Solid Waste Management regarding post earthquake inspection dated July 8, 1992.

Dewey, J.W., Reagor, B.G., Dengler, L., Moley, K., "Intensity Distribution and Isoseismal Maps for the Northridge, California, Earthquake of January 17, 1994", United States Department of the Interior Geological Survey National Earthquake Information Center, pages 2-4.

Finch, M.O., May 24, 1995, personal communication effects of the Loma Prieta earthquake on the Redwood, Zanker Road, Guadalupe, and Santa Cruz Landfills.

Hake and Cloud (1969). United States Earthquakes 1969, National Oceanic and Atmospheric Administration.

Johnson, M.E., Lundy, J., Lew, M., Ray, M.E., (1991). "Investigation of Sanitary Landfill Slope Performance During Strong Ground Motion from the Loma Prieta Earthquake of October 17, 1989," in Proceedings: Second International Conference on Recent Advances in Geotechnical Earthquake Engineering and Soil Dynamics, March 11-15, 1991, St. Louis Missouri, Paper No. LP27, page 1707.

Krinitzsky, E.L., Chang, F.K., Nuttli, O.W., (1988). "Magnitude-Related Earthquake Ground Motions" in Bulletin of the Association of Engineering Geologists, Volume XXV, No. 4, 1988, pages 399-423.

Landtech (1995). "Toyon Canyon Landfill; Northridge Earthquake Update" prepared for the California Integrated Waste Management Board 2p.

References

O'Rourke, J.T., 1973, "Geological Investigation Redwood Sanitary Landfill, Inc. Marin County, California" Page 11.

Orr, W. R., and Finch M. O. "Solid Waste Landfill Performance During the Loma Prieta Earthquake," in Geotechnics of Waste Fills--Theory and Practice, ASTM STP 1070, Arvid Landva, G. David Knowles, editors, American Society for Testing and Materials, Philadelphia, 1990, p.22-30.

Purcell, Rhoades and Associates, 1986, Engineering Report Review in Conjunction With The 5-Year Permit Review, Sunshine Canyon Sanitary Landfill, Prepared for Browning-Ferris Industries of California, Page 50.

EVALUATION OF SOLID WASTE LANDFILL PERFORMANCE DURING THE NORTHRIDGE EARTHQUAKE

by Anthony J. Augello[1] S.M. ASCE, Neven Matasovic[2] A.M. ASCE, Jonathan D. Bray[1] M. ASCE, Edward Kavazanjian, Jr.[2] M. ASCE, and Raymond B. Seed[1] M. ASCE

ABSTRACT

The lack of well documented case histories of landfill performance during earthquakes has made it difficult to calibrate our analytical tools and understanding of the earthquake resistance of these engineered structures. Moreover, it has not been possible to use case histories of landfill performance during earthquakes to demonstrate regulatory compliance. The 1994 Northridge earthquake provides an excellent opportunity to document the seismic performance of solid waste fills. In this paper, seven case records of landfill performance during the Northridge Earthquake are described. Back-analyses of the performance of four unlined and three geosynthetic-lined waste units are used to characterize the dynamic strength of the waste fill materials and geosynthetic liner interfaces. The results of these analyses indicate that the dynamic strength of waste fill is higher than the shear strength commonly assumed in static stability analyses and that the dynamic strengths of liner interfaces are generally consistent with values commonly used in practice.

---

[1] Research Assistant, Assistant Professor and Professor, University of California, Berkeley, CA 94720
[2] Assistant Project Engineer and Associate, GeoSyntec Consultants, Huntington Beach, CA 946247

INTRODUCTION

Recent earthquake events, such as the 1987 Whittier Narrows, 1989 Loma Prieta and 1994 Northridge Earthquakes, have provided excellent opportunities to document the seismic performance of waste fills. The 1987 M_w = 6.0 Whittier Narrows Earthquake provided some of the first observational data on landfill seismic performance. Performance information on six solid waste landfills, two of which were within the zone of strong shaking, is available. None of these landfills were equipped with a geosynthetic or clay liner system at the time of the earthquake. The most significant damage observed in this event, was cracking in the cover soils at the Operating Industries, Inc. (OII) landfill, which is located 5 km from the earthquake epicenter (Woodward-Clyde 1988). The OII site was subjected to an estimated free-field maximum horizontal acceleration (MHA) of between 0.3 g to 0.4 g during this earthquake. No evidence of slope instability was noted. This event provided motivation for the installation of a pair of strong motion recording instruments at the OII landfill. At the Puente Hills landfill, located approximately the same distance from the 1987 Whittier Narrows Earthquake epicenter as the OII landfill, no damage was reported (Earth Technology Corp. 1988). Coduto and Huitric (1990) reported that no noticeable deformations were recorded in inclinometers and settlement monuments at the Mission Canyon landfill, located outside the zone of strong shaking, approximately 22 km west of the earthquake epicenter.

The 1989 M_w = 6.9 Loma Prieta Earthquake also provided useful observational data on seismic landfill performance. Nineteen landfills that experienced free-field MHAs on the order of 0.05 g to 0.4 g were inspected soon after the earthquake. The most common form of damage was cracking of the soil cover at waste fill/native ground contacts and at changes in landfill geometry (Buranek and Prasad 1991, Johnson et al. 1991, and Orr and Finch 1990). These cracks were typically on the order of 25 to 75 mm wide. From a stability perspective, even in the epicentral region where the side slopes of landfills had inclinations ranging from 2H:1V to 3H:1V and were up to 76 m high, solid waste landfills performed well. However, a number of landfill gas collection systems were affected by the loss of power and above-ground pipe breaks. Sharma and Goyal (1991) described the performance of the West Contra Costa landfill, which is sited atop a 12 to 18 m-thick deposit of young San Francisco Bay Mud at a distance of over 100 km from the epicenter. Amplification of the MHA by both the Bay Mud foundation soil and the overlying waste mass was predicted at the site by these investigators for this event. Inclinometer measurements indicated that no significant deformations were experienced by the landfill as a result of the earthquake.

In the M_w = 6.7 1994 Northridge Earthquake, twenty-two landfills within 100 km of the epicenter were subjected to ground motions in excess of 0.05 g (Matasovic et al. 1995). This paper presents the results of an evaluation of the earthquake resistance of six of these facilities, including four unlined waste units (Operating Industries, Inc. (OII), Sunshine Canyon, Lopez Canyon, and Toyon Canyon) and three geosynthetic-lined waste units (Chiquita Canyon, Lopez Canyon, and Bradley Avenue) during the Northridge Earthquake. This investigation was undertaken to characterize the dynamic response of solid waste and geosynthetic lining systems. The seismic performance of these landfills, along with the primary characteristics of each landfill including landfill geometry and waste fill characteristics, are presented herein. The results of back-analyses to characterize the dynamic strength of solid waste materials and geosynthetic interfaces are discussed, and a number of pertinent observations are made.

PERFORMANCE DURING THE NORTHRIDGE EARTHQUAKE

General

The 1994 Northridge Earthquake provides a good set of observational data on the response of landfills to strong levels of shaking (Augello et al. 1995, EERI 1995, Matasovic et al. 1995, Stewart et al. 1994). There are numerous active, inactive and closed landfills within 100 km of the epicenter of the Northridge Earthquake. Figure 1 shows the location of 22 solid waste landfills which experienced significant levels of shaking (i.e. free-field MHA > 0.05g). At 16 of these sites, the free-field rock MHA at the base of the landfill was estimated to be in excess of 0.24 g, and at six of these sites, the free-field rock MHA was estimated to be in excess of 0.38 g.

The closest distance to the zone of energy release, estimated free-field rock MHA, and damage category for the 22 landfills shown in Figure 1 are provided in Table 1. The closest distance to the zone of energy release refers to the distance from the "effective" fault plane as interpreted by Wald and Heaton (1994) to the approximate geometric center of the landfill. The free-field rock MHA was estimated as the mean value from the Idriss (1991) rock attenuation relationship for a $M_w = 6.7$ reverse fault event. This attenuation relationship has been shown to fit the Northridge Earthquake recorded rock data well, however, it tended to underpredict the MHA slightly in the near-field (i.e. < 20 km, Chang et al. 1996). At the OII landfill, the free-field peak acceleration at the base of the landfill was also obtained from strong motion instrumentation (Hushmand Associates 1994). The damage categorization used was developed by Matasovic et al. (1995), and it is based upon impairment to the waste containment system and requirements for post-earthquake repair.

The Northridge Earthquake provides the first opportunity to examine the seismic performance of geosynthetic-lined landfills. Of the 22 landfills, eight have composite liner systems containing geosynthetic liner elements (Table 1). The most commonly observed damage patterns at the 22 landfills was cracking of the soil cover at waste fill/native ground contacts and at changes in landfill geometry. This pattern of damage is consistent with the damage observed after the Loma Prieta Earthquake. This cracking was most pronounced at the Sunshine Canyon landfill, the landfill closest to the zone of energy release. Two localized tear areas in the geomembrane side slope liners, one 4 m in length and the other about 27 m in length, occurred at the Chiquita Canyon landfill (EMCON Associates 1994). Two other geosynthetic-lined landfills (Bradley Avenue and Lopez Canyon landfills) at similar distances from the zone of energy release suffered no apparent damage to their geosynthetic liner systems. However, these two landfills did suffer moderate damage evidenced by cracking in the cover soil at waste fill/native ground transitions, breaking of gas extraction systems header lines, and loss of power to the gas collection system. Similar to the Loma Prieta Earthquake, the Northridge Earthquake damaged above ground pipes and caused power losses to landfill gas extraction systems. The strong motion records obtained at the OII landfill in the Northridge Earthquake provide a valuable opportunity to calibrate analytical tools and to back-calculate dynamic waste fill properties. The characteristics and seismic performance of the six key landfills impacted by the Northridge Earthquake are discussed next, followed by an engineering analysis of their performance.

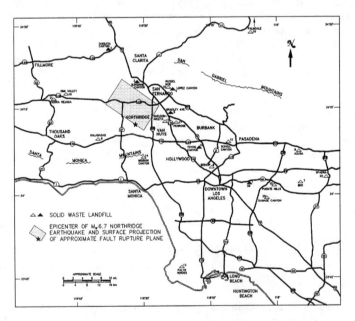

Figure 1.   Major solid waste landfills within 100 km of the January 17, 1994 Northridge Earthquake Epicenter (after Matasovic et al 1995)

Table 1. Seismic performance of solid waste landfills during the 1994 Northridge Earthquake (after Matasovic et al. 1995)

SOLID WASTE LANDFILL	TYPE OF LANDFILL	ENGINEERED CONTAINMENT SYSTEM	DISTANCE FROM ZONE OF ENERGY RELEASE (km)	ESTIMATED ROCK PEAK HORIZONTAL ACCELERATION (g)	DAMAGE CATEGORY (I - V)	DAMAGED ELEMENT
1. OII	Gravel Pit Fill	None	43	0.10 (0.24)	Minor Damage (II)	Cover Soil
2. Toyon Canyon	Canyon Fill	LCRS; Subsurface Barrier	22.2	0.21	Minor Damage (II)	Cover Soil; Gas Collection Header
3. Sunshine Canyon	Side Hill Fill	None	7	0.46	Moderate Damage (III	Cover Soil
4. Lopez Canyon	Canyon Fill	None (Disposal Areas A,B, AB+); Geosynthetic Liner System & LCRS (Disposal Area C)	8.4	0.42	Moderate Damage (III	Cover Soil; Gas System
5. Chiquita Canyon	Canyon Fill	None (Primary Landfill) Geosynthetic Liner System & LCRS (All other Canyons)	12.2	0.33	Significant Damage (IV)	Cover Soil; Geomembrane Tears
6. Bradley Avenue	Gravel Pit Fill	None (East); Clay Liner (West);Geosynthetic Liner System & LCRS (West Extension)	10.8	0.36	Moderate Damage (III	Cover Soil
7. Russel Moe	Canyon Fill	None	7.8	0.43	Moderate Damage (III	Cover Soil
8. Sheldon-Arleta	Gravel Pit Fill	None	10.7	0.36	Minor Damage (II)	Cover Soil; Gas Collection Headers
9. Penrose	Gravel Pit Fill	None	12.3	0.33		
10. Mission Canyon	Canyon Fill	None	18.4	0.25	No Damage (I)	None
11. Simi Valley	Canyon Fill	Geosynthetic Liner System & LCRS; Compacted Soil	22.3	0.21	Minor Damage (II)	Cover Soil; Gas System; Leachate Pump
12. Terra Rejada	Canyon Fill	None	22.4	0.21	Minor Damage (II)	Cover Soil
13. Calabasas	Canyon Fill	Geosynthetic Liner System & LCRS (Cell P)	23.1	0.20	Moderate Damage (III	Gas System; Cover Soil
14. Scholl Canyon	Canyon Fill	None	28.4	0.16	Moderate Damage (III	Cover Soil
15. Bishop Canyon	Canyon Fill	None	30.7	0.15	Little Damage (I)	Cover Soil
16. Palmdale	Area Fill	None	41.1	0.11	Minor Damage (II)	Cover Soil
17. Puente Hills	Side Hill Fill; Canyon Fill	Geosynthetic Liner System & LCRS (Canyon 9)	49.7	0.09	No Damage (I)	None
18. Palos Verdes	Canyon Fill	None	50.8	0.08	No Damage (I)	None
19. Azusa	Gravel Pit Fill	Geosynthetic Liner System & LCRS (Partial Coverage)	51.7	0.08	No Damage (I)	None
20. Savage Canyon	Canyon Fill	None	52.8	0.08	No Damage (I)	None
21. Spadra	Canyon Fill	Geosynthetic Liner System (Western Half) & LCRS	55.1	0.07	No Damage (I)	None
22. BKK	Canyon Fill	Compacted Soil (Haz. Waste Unit)	57.2	0.07	No Damage (I)	None

DAMAGE CATEGORY: DESCRIPTION:
V. Major Damage General instability with significant deformations. Integrity of waste containment system jeopardized.
IV. Significant Damage Waste containment system impaired, but no release of contaminants. Damage cannot be repaired within 48 hours. Specialty contractor needed to repair the damage.
III. Moderate Damage Damage repaired by landfill staff within 48 hours. No compromise of the waste containment system integrity.
II. Minor Damage Damage repaired without interruption to regular landfill operations.
I. Little or No Damage No damage or slight damage but no immediate repair needed.

Operating Industries, Inc. Landfill

The OII landfill is located in the City of Monterey Park, approximately 16 km east of downtown Los Angeles (Fig. 1). The landfill is located within the Montebello Hills which are part of a series of low lying hills (Repetto, Montebello and Puente) separating the San Gabriel Valley from the Los Angeles Basin. The landfill overlies the Lakewood/San Pedro Formations of the Pleistocene period and the Pico Unit of the Pliocene period. It is likely that sands and gravels of the Lakewood/San Pedro formation and recent alluvium were removed from beneath the landfill by quarrying operations before placement of trash at the site. The Pico Unit at the site consists primarily of a thick siltstone bed. The siltstone ranges from a clayey siltstone to a very fine grained sandy siltstone and silty fine-grained sandstone. Occasional channel deposits, consisting of sandstones, conglomeritic sandstones and conglomerate rocks incised into the Pico Unit, underlie the site. These rocks are considered very dense and consolidated, but poorly indurated.

The OII landfill is approximately 76 hectares in size, separated into two parcels by the Pomona Freeway. The south parcel, about 59 hectares in size, is the primary landfill unit (Fig. 2). Prior to 1946, the site was operated as a sand and gravel quarry. Waste disposal operations began at the site in 1948. The entire landfill was constructed without a liner system. Prior to 1954, the landfill received only solid waste. Starting in 1954, the landfill was permitted to receive liquid wastes. In 1976, disposal of liquid hazardous waste was restricted to a 13 hectare area in the southwest portion of the site. In 1983, disposal of liquid hazardous wastes ceased, and in October of 1984, landfilling operations ceased. The landfill is currently awaiting closure as a Superfund site. In 1987, the landfill became relatively well instrumented with the installation of survey monuments, slope inclinometers and a pair of strong motion recording stations, one on the top deck and one adjacent to the toe of the eastern end of the landfill (Seigel et al. 1990). The elevation of the top deck of the south parcel ranged from El +192 m to El +195 m at the time of closure. Interim cover soils consist mainly of silty clay to silty sand and range in thickness from about 0.3 m to as much as 9 m (Mundy et al. 1995). The maximum thickness of the waste at the site is approximately 100 m with sides slopes ranging in height from 20 to 70 m above the adjacent ground. These slopes range from as steep as 1.3H:1V to 3H:1V. Figure 2 also shows cross sections through the north central slope of the south parcel (H-H') with a overall slope inclination of 1.9H:1V and through the eastern end of the landfill (A-A') where the strong motion recording stations are located.

CH2M Hill (1988) estimated trash volumes and placement quantities versus time based upon the contours of the trash prism and site history. Approximately 7.5 percent of the refuse was placed prior to 1956, the bulk of the waste was placed from 1956 through 1974 ($\cong$ 80 %) and the remaining 12.5 percent placed between 1974 and 1984. Based upon the development of the site, the oldest refuse is located in the north central portion of the south parcel. Subsequent filling proceeded to the south, east and finally to the west. The youngest wastes disposed of at the site are probably found from El +165 m to El +192 m across the entire site. Landfill records indicate that a majority of the liquid wastes (1220 million liters) were received after 1976, and that most of liquids were disposed of in the southwest portion of the south parcel (CH2M Hill, 1988).

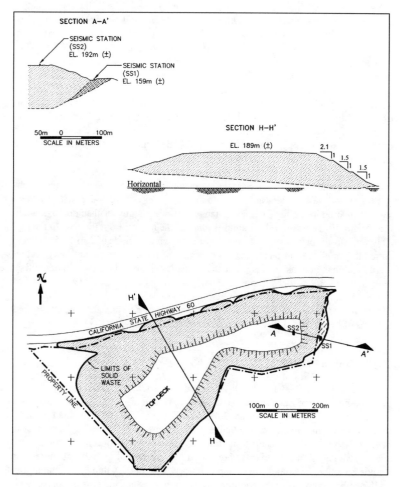

Figure 2. Plan View and cross section through the OII landfill (after CDM Federal Programs 1995)

The OII landfill is about 43 km from the zone of energy release of the Northridge Earthquake. The base seismic station recorded a MHA of 0.24g (east-west or longitudinal direction) and the top seismic station recorded a MHA of 0.25g (east-west direction) during the Northridge earthquake. The Idriss (1991) rock attenuation relationship would predict a median horizontal acceleration of 0.1 g and a median plus two standard deviations MHA of 0.25g at the OII landfill for a $M_w = 6.7$ event. Hence, the MHA value recorded at the base of the OII landfill falls just below the median plus two standard deviation value. A recent soil boring and geophysical testing program has determined that the base seismic station is sited upon approximately 26 m of artificial fill (Pacific Engineering and Analysis 1995). After the earthquake, minor cracking was observed at a number of locations on the faces of slopes, primarily but not exclusively, at or near to the berm roads. The cracks were generally on the order of 50 to 150 mm or less at their widest point. From observations made during trenching after the earthquake, the cracks did not appear to extend fully through the soil cover system into the underlying waste. Instrumentation data collected after the earthquake indicated that the landfill did not experience significant horizontal or vertical deformations as a result of the earthquake (CDM Federal Programs 1995).

Acceleration-time histories and acceleration response spectra for the north-south and east-west motions are shown in Figure 3. For both recorded motions, there was attenuation in the high frequency range. At periods greater than approximately 0.6 seconds, these records show amplification of the motion from the base to the top. This was most pronounced in the east-west direction at periods of 1 to 1.25 seconds, where the amplification factor was on the order of three. The spectral acceleration amplification functions indicate that the fundamental period of the OII landfill is approximately 1.2 seconds in both the north-south and east-west directions and that the landfill responds primarily in its first mode (Stewart et al. 1994). This observation is consistent with those made from previous earthquakes which induced motions of lower intensity at the landfill (Hushmand Associates 1994).

Toyon Canyon Landfill

The Toyon Canyon landfill is located in the City of Los Angeles' Griffith Park on the northern flank of the Santa Monica Mountains, which separates the Los Angeles Basin from the San Fernando Valley (Fig. 1). The site is situated within a northwest-trending, overturned anticline offset by a northeast trending fault that is covered by the landfill. The site is primarily underlain by Miocene conglomerates and sandstones of the Hollycrest Formation. The exposed bedrock at the site is weathered to a maximum depth of approximately 4 to 5 m. A thin layer of alluvium and/or colluvium overlies the bedrock. The alluvium, which consists of coarse sand, gravel and silt, ranges in thickness from about 0.45 m along the canyon slopes to approximately 6 m at the mouth of the canyon (IT Corp. 1994).

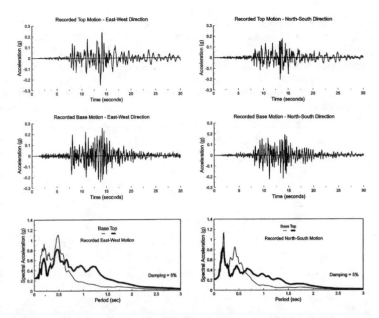

Figure 3. Acceleration time-histories and response spectra for the recorded east-west and north-south motions at the OII landfill

The landfill, which is approximately 36 hectares in size, began operations in 1957 in a steeply sloped canyon as a municipal solid waste landfill. Landfilling progressed from the lower elevation of +183 m to a maximum elevation of +314 m by the cut and cover method. Soil cover for daily operations was obtained from excavations within the site or along the canyon slopes. The landfill stopped receiving wastes on a daily basis in 1985 and discontinued operations in February 1986. It currently has an interim cover and is awaiting final closure. The landfill has a relatively flat top deck covering approximately 16 hectares with a refuse slope across the mouth of the canyon ranging from 2H:1V to 4H:1V. The existing soil cover along the top deck ranges from 1 m thick in the center to about 2 m thick at the crest of the slope to about 3.65 m thick at the back end of the top deck. The refuse slope, which is 130 m high, was constructed with eleven benches. The refuse slopes' cover soil is approximately 1.2 m thick. Figure 4 shows a plan view and cross section through the Toyon Canyon Landfill with an overall slope inclination of about 3H:1V.

This landfill, operated by the City of Los Angeles, received approximately 14.5 million metric tons of municipal solid waste over its operational life at a rate of about 1500 metric tons/day from 1957 to 1967 and about 2430 metric tons/day from 1968 to 1985. The wastes received at the site consist primarily of paper products ($\cong$ 34 %) and leaves, grass, tree trimmings and wood ($\cong$ 37%). The maximum thickness of waste is approximately 79 m. Based upon topographical maps compiled from air photos, IT Corporation (1994) estimated settlements on the top deck ranging from approximately 4.5 m at the face of the slope to 7.3 m at the back of the landfill from 1985 through 1992.

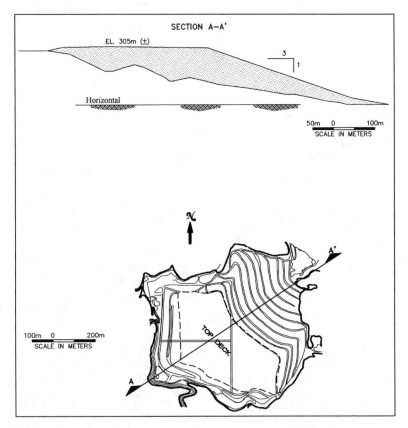

Figure 4. Plan view and cross section through the Toyon Canyon landfill (after IT Corp. 1994)

The Toyon Canyon landfill is about 22 km from the zone of energy release of the Northridge Earthquake. A nearby strong motion station at the Griffith Park Observatory recorded a MHA of 0.29 g. The Idriss (1991) rock attenuation relationship would predict a mean peak horizontal bedrock acceleration of 0.21 g at this site for a M_w = 6.7 event. After the earthquake, longitudinal cracks in the soil cover were noticed along the western portion of the top deck and along the front faces of the slopes and benches. Cracks were also observed in the cover soils at the waste fill/native ground interface along the eastern half of the landfill. Cracks in the cover soil were on the order of 25 mm wide and typically less than 50 mm wide. The landfill gas to energy recovery facility located at the site shut down as a result of a loss of power. In addition, a landfill gas collection header was sheared off and several breaks in the water supply pipeline to the cooling towers for the energy recovery facility were discovered after the earthquake.

Sunshine Canyon Landfill

The Sunshine Canyon landfill is located in the city of Sylmar just southwest of the Interstate Highway 5 / Highway 14 interchange at the eastern edge of the Santa Susana Mountain Range (Fig. 1). The landfill lies at the eastern nose of the Pico Anticline, an anticline that generally trends east-west along the crests of the Santa Susana Mountains. The bedrock at the site is the Pliocene and Miocene age Towsley Formation. The Towsley Formation lithology ranges from shale to boulder conglomerates, but at the site it is dominated by sandstone and siltstone. The coarser horizons are generally better cemented and form most of the cliffs in the area. The fine-grained horizons, particularly the interbedded shales and siltstones, are frequently internally contorted, presumably due to syndepositional slumping and loading.

This landfill, which is a side-hill fill, began operations in 1958 as a municipal solid waste sanitary landfill. It stopped receiving wastes in 1991 and is currently inactive, awaiting final closure. The landfill is 93 hectares in size and consists of two separate units, the main landfill unit and the north landfill unit (Fig. 5). The main landfill, which is unlined, has an interim soil cover approximately 2.5 m thick. The soil for both the daily and the intermediate covers was obtained from excavations within the site. At the main landfill, landfilling progressed from a lower elevation of +425 m to a maximum elevation of +595 m. The refuse received at the site consisted mainly of municipal solid waste, however, construction debris and sewage sludge were also accepted. The maximum thickness of refuse is about 106 m. Figure 5 shows a cross section through the main landfill unit, which has an average overall slope face inclination of 2.6H:1V.

The Sunshine Canyon landfill is located only 7 km from the zone of energy release. Strong motion stations in the area on recent alluvium recorded MHAs on the order of 0.9 g, but these values were influenced by site effects (Chang et al. 1996). MHAs at nearby rock sites ranged from 0.4 g to 0.45 g. The Idriss (1991) rock attenuation relationship predicts a mean peak bedrock acceleration on the order of

0.46 g at this site for a magnitude 6.7 event. Longitudinal cracks were observed at the top of the landfill along the interface with the native canyon walls. The cracks varied from less than 20 mm to as wide as 300 mm, with 150 to 300 mm of vertical offset in some areas. Figure 6 shows the cracking observed in the soil cover of the top deck at the western end of the landfill. Cracking was also observed at the edge of several benches along the face of the waste slope. These cracks were generally less than 20 mm wide. This cracking was not believed to represent an overall instability.

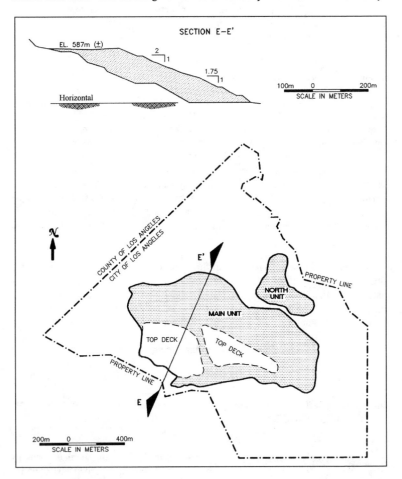

Figure 5. Plan view and cross section through the Sunshine Canyon landfill (after GeoSyntec Consultants 1995)

Figure 6. Crack along top deck at western end of the Sunshine Canyon Landfill (photo courtesy of Calif. EPA, Integrated Waste Management Board)

Lopez Canyon Landfill

The Lopez Canyon landfill is located in the foothill region of the San Gabriel Mountains, approximately 50 km northwest of downtown Los Angeles (Fig. 1). The site is underlain by bedrock of the Tertiary Modelo and Towsley Formations, and the Tertiary-Quaternary Saugus Formation. Quaternary terrace deposits are present locally near the southeastern boundary of the property. Holocene and Quaternary alluvium line drainage channels and canyon floors.

The landfill, which has a total capacity of 16.9 million metric tons of refuse, began operations in 1975 as a municipal solid waste landfill. This landfill consists of four disposal area designated: Areas A, B, AB+, and C (Fig. 7). Disposal areas A, B, and AB+ are no longer accepting waste and are awaiting final closure. Disposal areas A and B, which are unlined and cover about 30 hectares of land, were the initial landfill units. Figure 8 shows the most critical cross section, in terms of stability, through Area A with a slope height of 51 m and an overall inclination of 2.5H:1V. Disposal Area C, which covers about 15 hectares of land, is the only active cell of the landfill. Canyon C's native side slopes are up to 90 m high and have been graded to provide slopes of between 1H:1V to 1.5H:1V, with 5 m wide benches every 12 m in height. The base and side slopes were lined with a Subtitle D composite liner system (Derian et al. 1993). The base liner system conforms to the prescriptive requirements of Subtitle D, consisting of a 0.3 m thick leachate collection layer overlying a composite liner. The composite liner consists of a 0.6 m low permeability native soil/bentonite admix (4% bentonite by weight) covered by a 2.0 mm (80 mil)-thick textured high density polyethylene(HDPE) geomembrane. There

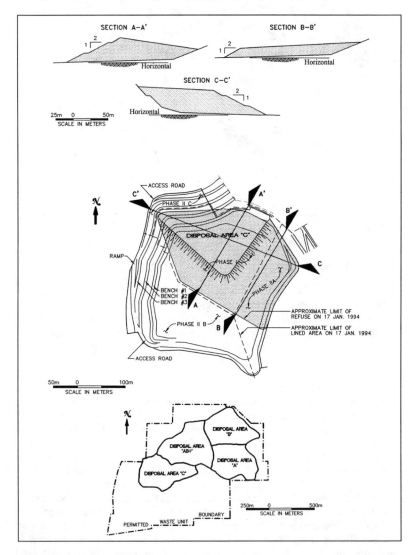

Figure 7. Plan View and cross sections, Phase I, Disposal Area C, Lopez Canyon landfill (after GeoSyntec Consultants 1994)

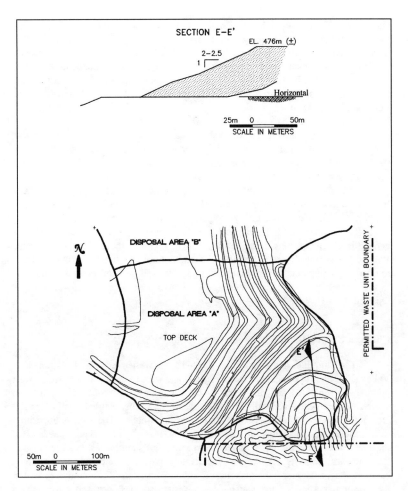

Figure 8. Plan view and cross section, Disposal Area A, Lopez Canyon landfill (after GeoSyntec Consultants 1994)

is a 545 g/m^2 non-woven cushion geotextile between the geomembrane and the leachate collection layer and a 410 g/m^2 filter geotextile overlain by 0.6 m of protective soil on top of the leachate collection layer. The side slope liner system is an alternative liner system designed in conformance with the performance standards of Subtitle D. The side slope liner consists of, from top to bottom, a 0.6 m thick protective soil layer, a 410 g/m^2 filter geotextile, a geonet drainage layer, a 2.0 mm-thick HDPE single-sided textured geomembrane (textured side down), a geosynthetic clay liner (GCL) and an air-sprayed slope veneer of concrete averaging 75 - 100 mm thick, reinforced with 17-gauge wire hexagonal netting (Derian et al. 1993). Construction of the liner system at Disposal Area C was divided into two phases. Phase I (Fig. 7), which covers approximately 8 hectares, was completed in May 1993 and was filled to a height of about 30 m at the time of the Northridge Earthquake. Figure 7 shows several cross sections through Phase I of Disposal Area C at the Lopez Canyon landfill.

The Lopez Canyon landfill is about 8 km from the zone of energy release. Rock stations in the area recorded peak ground accelerations on the order of 0.4 g to 0.45 g. The Idriss (1991) rock attenuation relationship predicts a mean peak ground acceleration of 0.42 g at this site for a magnitude 6.7 event. Minor cracking was observed in the interim soil cover at the interface between the older unlined waste fills and natural canyon slopes. The cracks in this area were minor, typically being 20-30 mm wide. There was no sign of permanent relative displacement between the waste fill and liner system in Disposal Area C. A small tear (150 to 200 mm) in the geotextile overlying the side slope liner was reported by the California Integrated Waste Management Board (CIWMB). Subsequent investigations attributed this tear to operating equipment (GeoSyntec 1994). The landfill also suffered minor damage to the surface gas extraction system (broken gas header lines), which was quickly repaired.

Chiquita Canyon Landfill

The Chiquita Canyon landfill is located at the western edge of the Santa Clara Valley (Fig. 1). Topography to the north, east and west is characterized by east-west orientated steep-sided canyons with slopes that typically approach 1H:1V and in some cases are nearly vertical. To the south, the terrain is relatively flat. This flat area defines the limits of the Santa Clara River flood plain. The Saugus Formation (lower Pleistocene) accounts for approximately 80 percent of the exposed bedrock at the site. This formation, which ranges from rolling hills in the west to steep cliffs in the southeast, consists of poorly to moderately indurated interbedded sandstone, conglomerate, siltstone, and mudstone. The Pico Formation (Pliocene) is exposed in the northern portion of the site as steep, resistant dip slopes and cliffs. This formation consists of cemented sandstone with interbedded conglomerate and siltstone. This formation is more resistant to weathering than the Saugus Formation. The colluvium and alluvium, which range in thickness from 3.6 to 10 m on the canyon floor, consist of silty sand and sand with some gravel, silt and clay. Some of this colluvium/alluvium has been excavated during landfilling operations for use as daily and intermediate cover (EMCON Associates 1991).

The Chiquita Canyon landfill, which lies on approximately 200 hectares of land, consists of five landfills designated: the Primary Canyon landfill and Canyons

A, B, C and D. The landfill began operations in 1972 with the opening of the Primary Canyon landfill. Currently, the Primary Canyon and Canyon B landfills are inactive and awaiting final closure. The Primary Canyon, which is about 22 hectares in size, operated from 1972 through 1988. The Canyon B landfill (6 hectares) was constructed in 1988 and stopped receiving wastes in July of 1989. Canyons A (7.7 hectares) and D (4 hectares) are partially filled and only used for landfilling during wet weather. Canyon C, which consists of two cells, is the active area of the landfill. The geosynthetic liner system for Cell I was completed in 1991. This cell, which is approximately 10.6 hectares in size, is actively receiving waste fill. Cell II is currently in the final phase of design (EMCON Associates 1994).

Two tears in the geomembrane side slope liner occurred in Canyons C and D. Therefore, this paper focuses on these two areas of the site (Fig. 9). At the Canyon C landfill, the free face of the active waste fill has a slope of about 2H:1V. The side slopes cut into the canyon wall vary from 1.5H:1V to 2H:1V. These slopes are covered with a 1.5 mm (60mil)-thick smooth HDPE geomembrane liner. The base of the landfill is fitted with a composite liner consisting of a 0.6 m thick bentonite-soil admix underlying a single-sided textured HDPE geomembrane (textured side down). The base of the landfill has a leachate collection layer which consists of a network of polyvinyl chloride (PVC) pipes within a 0.3 m-thick gravel layer. The maximum refuse depth at the time of the earthquake was approximately 30 m. Figure 9 shows a plan view and cross section through the Canyon C landfill in the area of the geomembrane liner tear. At the Canyon D landfill, the cut side slopes are approximately 30 m high with an inclination of 2.5H:1V. The base of this area is lined with a 0.3 m-thick soil liner consisting of a mixture of alluvium and 9 percent bentonite (by dry weight). The leachate collection system consists of a 150 mm sand layer over the base soil liner. The side slopes are lined with a 1.5 mm (60 mil)-thick smooth HDPE geomembrane placed directly over subgrade. The geomembrane is covered with 0.6 m of protective soil cover. Final refuse slopes are maintained at an inclination of approximately 3H:1V. Figure 9 shows a plan view and cross section through the Canyon D landfill where the tear in the geomembrane liner occurred.

Chiquita Canyon is a municipal solid waste landfill which receives municipal solid waste, industrial and demolition debris, and sewage sludge. The majority of the waste received at the site (76 %) is residential and commercial municipal solid waste. Industrial and demolition debris comprise about 12 % of the waste stream at the site. The landfill is permitted to receive sewage sludge with a solids content of greater than 50%. Sewage sludge comprises approximately 3 % of the waste stream. The waste stream is relatively free of all recyclable materials, as much of the municipal solid waste comes from transfer stations which conduct active recycling programs (EMCON Associates 1991). The refuse deposited in Canyons C and D is relatively recent in age, with most of the waste being placed in the last 8 years.

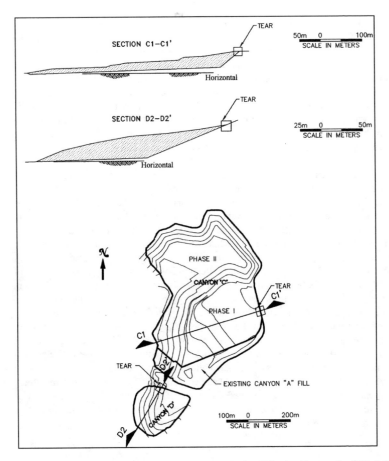

Figure 9. Plan view and cross sections through the Chiquita Canyon landfill (after EMCON Associates 1994)

Significant damage occurred at the Chiquita Canyon landfill as a result of the Northridge Earthquake. This damage includes limited downslope movement of the waste, cracks in the soil cover system, tears in the geosynthetic liner system, and a temporary shut down of the gas removal system due to a loss of external power. This landfill is about 12 km from the zone of energy release. At nearby rock sites, recorded MHAs ranged from 0.2 g to 0.4 g (Chang et al. 1996). The Idriss (1991) rock attenuation relationship predicts a mean peak ground acceleration of 0.33 g at the Chiquita Canyon site for a M_w = 6.7 event. At the time of the earthquake, only Canyon C was accepting waste fill. After the earthquake, cracks in the soil cover were observed in all areas of the landfill. In Phase I, Canyon C, longitudinal cracks were observed at the top of the landfill along the interface between the landfill liner and the waste fill. The largest cracks were approximately 300 mm wide, with vertical offsets of 150 to 300 mm. A localized tear in the geomembrane was observed in one area of Canyon C. The tear, which occurred at the top of the slope near the anchor trench (where the largest static (pre-seismic) stresses in the HDPE liner would be expected due to side slope downdrag as the fill compressed and settled over time), was approximately 4 m long and 0.24 m wide. Figure 10 shows the tear in the Canyon C geomembrane liner. It appears that this tear initiated at the location of an extrusion welded patch along a longitudinal seam where a sample was removed for destructive testing (EMCON Associates 1994). Minor cracking was observed in cover soils of the Primary Canyon and Canyon B landfills. At Canyons A and D, cracks parallel to the top of the slope were observed in the soil cover. In Canyon A, typical cracks were on the order of 150 mm wide with approximately 130 mm of vertical offset. The cracks in Canyon D were somewhat more pronounced. These cracks were as wide as 300 mm, with 200 mm of vertical offset, exposing the landfill liner in some areas. In February 1994, during landfill gas monitoring, a second tear area in the geomembrane liner was found in Canyon D. This tear area was comprised of three parallel tears, each approximately 0.3 m wide with a total length of 27 m, and these tears also occurred at the top of the side slope near the anchor trench.

Figure 10. Tear in HDPE geomembrane liner system, Canyon C, Chiquita Canyon Landfill (photo courtesy of Calif. EPA, Integrated Waste Management Board)

Bradley Avenue Landfill

The Bradley Avenue landfill is a gravel pit fill located in the Sun Valley section of Los Angeles, within the Hansen Subarea of the San Fernando Valley (Fig. 1). The Hansen Subarea is an irregularly shaped area between the crest of Hansen Dam, Pacoima Hills, Verdugo Fault and part of the Verdugo Mountains below the La Tuna Canyon Debris Dam. The alluvium at the site was eroded from the San Gabriel and Verdugo Mountains and deposited by the Tujunga wash during the Quarternary Period. The alluvium overlies the pre-Cretaceous age crystalline and metamorphic basement complex rocks. The older alluvium consists of brownish to reddish-grey silty, subangular sand, cobbles and boulders. These deposits, which are more than 150 m deep, are crudely horizontally stratified, unweathered and contain less than 1 percent clay. The Holocene alluvial deposits are mainly accumulations of light grey subangular boulders, gravels and sands. The deposits are uncemented but are tightly packed and stand at slopes of 1H:1V or steeper. These deposits, which are approximately 15 to 23 m thick, have been removed by previous quarry operations.

The Bradley landfill is an 84.5 hectare, municipal solid waste landfill composed of three contiguous working areas: Bradley East (28.3 hectares); Bradley West (28.7 hectares); and Bradley West Extension (27.5 hectares) (Fig. 11). Mining conducted over several decades resulted in a steep-walled pit of a fairly uniform depth of 49 m. Bradley East, which is an unlined facility, began operations in 1959 and is currently awaiting final closure. A small area in the northern part of this area has been closed and currently houses a sorting and recycling facility. Bradley West was designed between 1975 and 1977 with a 150 mm thick clay liner and a leachate collection and removal system consisting of a 300 mm thick sand and gravel layer. This area, which began operations in the early 1980s, is currently inactive. The Bradley West Extension began operations in 1987 and was designed and constructed with a geosynthetic liner system and a leachate collection system. The leachate collection system consists of a 300 mm gravel drainage layer overlying a geotextile filter. The composite base liner consists of a 1.5 mm (60 mil) HDPE geomembrane overlying a 300 mm compacted clay layer. A compacted soil berm lined with a geomembrane running the entire length of the site separates the Bradley East landfill from the Bradley West and Bradley West Extension landfills. The Sump 6 area (West Extension), which was receiving waste at the time of the earthquake, was completed in 1990 and began operations in 1991. Figure 11 shows a cross section through the Sump 6 area of the Bradley Avenue landfill.

The Bradley Avenue landfill is about 11 km from the zone of energy release of the Northridge Earthquake. MHAs recorded at nearby rock stations ranged from 0.2 g to 0.4 g (Chang et al. 1996). The Idriss (1991) rock attenuation relationship predicts a mean peak bedrock acceleration of 0.36 g at this site for a magnitude 6.7 event. After the earthquake, cracks were observed at the contact between the native ground and the waste fill along the eastern side of the Bradley Avenue East and West landfills. These cracks showed up to 25 mm of vertical offset. These cracks occurred along the waste/geomembrane liner interface and may have been the result of limited downslope movement along this interface. No tears were observed in the geomembrane liner at this landfill, however, tears in the geotextile overlying the side slope liner were observed after the earthquake. These tears were reported to have initiated prior to the earthquake but increased in size as a result of the earthquake strong shaking (California EPA 1994).

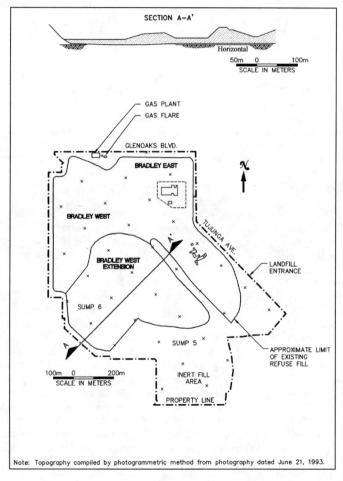

Figure 11. Plan view and cross section through the Bradley Avenue landfill (after Waste Management, Inc. 1991)

BACK-ANALYSIS OF LANDFILL PERFORMANCE

Analytical Procedures

Due to the difficulties of sampling and testing solid waste, observations and back-analysis of field performance provide valuable data on solid waste properties. The six key landfills of this study were back-analyzed to estimate static and dynamic properties of waste fill and landfill/liner interfaces. One-dimensional (1-D) wave propagation analyses using the computer programs SHAKE91 and D-MOD, and two-dimensional (2-D) dynamic finite element analyses using the computer program QUAD4M were performed to back-analyze representative landfill cross sections during the Northridge Earthquake. SHAKE91 (Idriss and Sun 1992) utilizes the equivalent linear elastic method in the frequency domain to model the nonlinear dynamic modulus and damping as a function of shear strain. D-MOD (Matasovic 1993) is a nonlinear time domain analysis using the Modified Konder-Zelasko constitutive model to describe the hysteretic behavior of solid waste during cyclic loading. QUAD4M (Hudson et al. 1994) utilizes the equivalent linear method with Rayleigh damping in a time domain plane strain, finite element analysis. These analytical procedures assume that no slip occurs along potential sliding planes, including base liner material interfaces. If slip occurs, the shear stresses at these interfaces will be less than the shear stresses presented herein.

Representative columns of waste and foundation materials for each of the landfill cross sections shown in the previous section were analyzed with the 1-D wave propagation programs. In addition, two 2-D cross sections (OII and Chiquita Canyon D) were modeled directly with QUAD4M. A typical unit weight profile for solid waste developed by Kavazanjian et al. (1995) was used in the analyses. This profile was developed based upon typical initial in-place densities and average unit weight and compressibility values reported in the literature. The initial static strength parameters of the waste fill were estimated from the bi-linear strength envelope developed for solid waste by Kavazanjian et al. (1995). These strength parameters were developed from back-analysis of the stability of several landfills along with values from laboratory and field testing reported in the literature. A Mohr-Coulomb strength envelop consisting of $\phi = 0$ with c = 24 kPa at normal stresses below 30 kPa and $\phi = 33°$ with c = 0 at higher normal stresses was used as the initial static strength parameters. The waste fill shear wave velocity profile developed by Kavazanjian et al. (1995) based upon reported in situ geophysical measurements and field measurements of ambient vibrations was also used in this study. This profile was varied by +30 % to provide an upper bound to the waste's shear wave velocity. The average modulus reduction and damping curves for solid waste proposed by Kavazanjian and Matasovic (1995) and those proposed by Vucetic and Dobry (1991) for a high plasticity (PI = 200) clay were used to represent the strain-dependent shear modulus and damping of solid waste. These curves were selected to provide reasonable variations of the shear strain dependent modulus reduction and damping characteristics of solid waste based upon ongoing studies of the recorded seismic response of the OII landfill.

The recorded base motion at the OII landfill was deconvolved using SHAKE91 to develop a rock outcrop motion for use in the dynamic analyses. As the OII landfill was the only landfill equipped with strong motion recording stations, two different rock motions were used at the other landfills: (a) motion from closest rock strong motion station (these nearby recorded rock input motions were scaled to the mean MHA predicted by the Idriss (1991) rock attenuation relationship for the actual distance from zone of energy release); and (b) synthetic rock motion generated by Pacific Engineering and Analysis employing a stochastic finite-fault model (Silva 1993, Schneider et al. 1993) incorporating the slip and crustal models of Wald and Heaton (1994).

Typical friction angles for geosynthetic liner interfaces in practice range from around 8° for smooth geomembrane interfaces to around 25° for textured geomembrane interfaces. Mitchell et al. (1990) performed a number of direct shear and pull-out tests to measure the static shear strength along geomembrane/geosynthetic interfaces. They found that the static shear strength along these interfaces was sensitive to the type of geosynthetic (smooth or textured), the quality of the geosynthetic (polished or unpolished) and wetting of the interface. The static interface friction angles along smooth geomembrane/geosynthetic interfaces ranged from 7° to 10°. Kavazanjian et al. (1991) performed shaking table and centrifuge model tests on layered geosynthetic interfaces and found little difference between static and dynamic interface shear resistance. Yegian and Lahlaf (1992) performed shaking table tests to measure the dynamic interface shear strength between geotextiles and geomembranes. They also found that the dynamic interface friction angle was approximately equal to that measured in static friction tests and that it was not significantly affected by the characteristics of the input base motion. Composite liner systems generally have a geomembrane/compacted clay interface. Seed and Boulanger (1992) performed a number of tests to measure the interface strengths between compacted clays and geomembranes. The frictional resistance along these interfaces was sensitive to the as-compacted water content of the clay and wetting along this interface. Peak friction angles along this interface ranged from 6° to as high as 30°, with residual friction angles as low as 5°, depending upon the compaction and wetting conditions.

The procedure used to estimate the dynamic strengths of waste fill and landfill/liner interfaces is as follows. A static slope stability analysis was first performed on each of the landfill cross sections shown previously to determine a potential critical slip surface using the program UTEXAS3 (Wright 1990). The 1-D dynamic analyses of representative waste columns were used to estimate the distribution of the maximum horizontal equivalent acceleration (MHEA) along this potential failure plane. This distribution of the MHEA was calculated for the two input rock motions at each site using the average shear wave velocity and unit weight curves combined with the average and upper bound modulus reduction and damping curves. The distribution of the MHEA weighted by the waste fill mass it acted on was used to estimate an average MHEA for that potential sliding mass. The average solid waste shear wave velocity and unit weight curves were used primarily, as they provided a conservative estimate of the average MHEA acting on a potential sliding mass. If the use of the average MHEA/g as the seismic coefficient in a pseudostatic stability analysis results in a factor of safety of one, the calculated seismically induced permanent displacement will be zero (Augello et al. 1995). Therefore, the average MHEA/g was multiplied by a factor corresponding to the amount of deformation observed at each landfill to estimate the seismic coefficient for use in the pseudo-static back-analysis of landfill stability. This factor, termed here the Observed Deformation Factor and labeled F, ranges from 0.7 for observed deformations of 50 mm or less to 0.4 for observed deformations of 300 mm or less. The factor, F, is dimensionless and essentially represents the ratio of the yield acceleration to the MHEA. The factor was developed based upon graphs presented in Augello et al. (1995). These factors are generally consistent with those recommended by Makdisi and Seed (1978) for earth embankments. This factor F is judgmental, and the near upper bound of the seismically induced permanent displacement versus ratio of the yield acceleration to MHEA (Augello et al. 1995) was used here based upon the reasoning that: (a) use of F=1.0 would be unconservative, as deformations were observed, and (b) use of mean or lower bound values from the curves would imply that these deformations appeared to be the result of shear displacement along relatively well-defined slip planes, which was not the case. Thus, the F values selected in this study are estimates intended to provide reasonably representative values specific to these back-analyses, and are not values that would necessarily be used for design.

Pseudostatic stability analyses were rerun on each of the landfill cross sections with the deformation calibrated seismic coefficient using UTEXAS3 to identify potential critical pseudostatic slip surfaces. If the critical pseudostatic slip surface differed from the critical static slip surface, the above procedure was repeated to refine the estimate of the deformation calibrated seismic coefficient for the new critical slip surface. This process was continued until convergence was achieved, as the critical pseudostatic slip surface is a function of the landfill geometry, material properties and seismic coefficient. After the seismic coefficient was estimated, a pseudostatic slope stability analysis was performed to back-calculate dynamic strengths along various potential sliding surfaces.

Figure 12 shows the finite element meshes used to represent OII cross section H-H and the Chiquita Canyon D landfill cross section. QUAD4M calculates directly the average horizontal equivalent acceleration-time history for a selected potential sliding mass. The HEA-time history is calculated as the ratio of the force induced by the earthquake over the weight of the sliding mass. The direction of the motion is important in these 2-D analyses. The MHEA in the downslope direction multiplied by the observed deformation factor was used as the seismic coefficient.

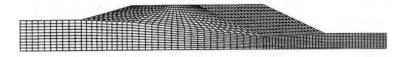

(a) OII landfill cross-section H-H (mesh length = 915 m)

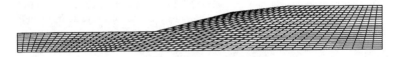

(b) Chiquita Canyon D landfill cross section (mesh length = 520 m)

Figure 12. Two-dimensional dynamic finite element meshes

The pseudostatic analyses should provide reasonable estimates of the dynamic strength of waste fill if the seismic coefficients are representative of the damage observed after the earthquake. Given the uncertainty with the input rock motions and waste fill dynamic properties, these strength estimates should be regarded as preliminary. Moreover, these back-analyses include assumptions of slip along a well defined failure plane and rigid movement of the sliding mass. It is recognized that although, slip may be concentrated along liner interfaces, this is unlikely within the waste fill where movements would occur over a more widely distributed shear zone . Nevertheless, these back-analyses are believed to be useful in providing approximate estimates of dynamic strengths for practice, because many engineering calculations made for the design of these types of facilities include these same assumptions.

Results

In the initial series of static stability analyses performed for this study, the solid waste fill's static friction angles were back-calculated assuming safety factors of 1.0, 1.1 and 1.2. As all six landfills encompassed in this study have no reported static stability problems, 1.2 is considered a reasonable lower bound on the existing static factor of safety. At unlined waste units, the critical sliding surface passed entirely through waste fill. For two geosynthetically lined waste units (Lopez Canyon Area C, Section A-A' and Chiquita Canyon C), the critical static slip surface passed partially along the liner surface and partially through waste fill. Stability along these potential sliding masses is affected by the liner interface shear strength as well as the waste fill shear strength. Typical friction angles for geosynthetic interfaces in practice range from approximately 8° for smooth geomembranes to 25° for textured geomembranes (Mitchell et al. 1990). The Lopez Canyon Area C landfill is lined with a double sided textured HDPE geomembrane along the base and a textured/smooth HDPE geomembrane along the side slopes, and the Chiquita Canyon C landfill is lined with a textured/smooth HDPE along the base and a smooth HDPE along the side slopes. The back-calculated friction angles for waste fill, along with the characteristics of the analyzed landfill slopes, are shown in Table 2. For the static analyses, the conservatively back-calculated solid waste friction angle ranged from 19° to 39° for a safety factor of 1.2. These values are on average lower than the previously cited value of 33° for solid waste (Kavazanjian et al., 1995); this is probably because a safety factor of 1.2 is a conservative assumption for these apparently stable landfill slopes.

Table 2. Static back-analysis of existing landfill slopes

	Average Slope		Maximum Slope		Waste Strength, ϕ (degrees)		
Landfill	Height (m)	Slope (H:V)	Height (m)	Slope (H:V)	FS = 1.0	FS = 1.1	FS = 1.2
OII Cross Section H-H'	75	1.9:1	27	1.4:1	29	32	35
Toyon Canyon	120	3:1	8	2:1	19	21	23
Sunshine Canyon	195	2.6:1	24	1.3:1	22	24	27
Lopez Canyon A Cross Section E-E'	51	2.5:1	15	2.3:1	19	21	23
Lopez Canyon C Cross Section A-A'	34	2.3:1	6.7	1.8:1	15	17	19

Table 3 shows the results of the pseudostatic stability analyses. The seismic coefficients estimated from the 2-D dynamic analyses are consistent with those values estimated from the 1-D analyses. The back-calculated dynamic friction angles at the four unlined landfills ranged from 27° to 52°. The larger friction angles were calculated from analyses utilizing higher estimates of the seismic coefficient, and these strength values may not be conservative. A reasonably conservative range of back-calculated dynamic friction angles at the four unlined landfills appears to be on the order of 30° to 40°. Based upon the results of these analyses coupled with the observations of landfill performance in locations close to the zone of energy release (< 12 km), these estimates for the dynamic friction angles at these landfills are greater than the lower bound static friction angles estimated using a static safety factor of 1.2.

Table 3. Pseudostatic back-analysis of existing landfill slopes

LANDFILL	ESTIMATED SEISMIC COEFFICIENT	ANALYSIS ASSUMPTIONS	OBSERVED DEFORMATION FACTOR	CLAY LINER STRENGTH, S_u ($\phi = 0$) (kPa)	GEOSYNTHETIC LINER STRENGTH ϕ (degrees) (c=0)	WASTE STRENGTH, ϕ (degrees) (c=0)
OII Cross Section H - H	0.11	1, I, A_{sym}, WS	0.7			39
	0.16	1, I, A_{rec}, WS	0.7			42
	0.18	2, I,II, A_{rec}, WS	0.7			43
	0.23	1, II, A_{rec}, WS	0.7			44
Toyon Canyon	0.13	1, I, A_{sym}, WS	0.7			27
	0.20	1, I, A_{rec}, WS	0.7			30
	0.21	1, II, A_{sym}, WS	0.7			31
	0.32	1, II, A_{rec}, WS	0.7			36
Sunshine Canyon	0.18	1, I, A_{rec}, WS	0.4			33
	0.22	1, I, A_{rec}, WS	0.4			35
	0.29	1, II, A_{rec}, WS	0.4			39
	0.56	1, II, A_{sym}, WS	0.4			52
Lopez Area A	0.28	1, I, A_{sym}, WS	0.7			36
	0.43	1, I, A_{rec}, WS	0.7			45
Lopez Area C Cross Section A - A'	0.19	1, I, A_{sym}, A_{rec}, LS	0.7		14	
	0.28	1, II, A_{sym}, LS	0.7		19	
	0.46	1, II, A_{rec}, LS	0.7		28	
			0.7			
Lopez Area C Cross Section B - B'	0.23	1, I, A_{sym}, LS	0.7		18	
	0.25	1, I, A_{rec}, LS	0.7		19	
	0.36	1, I, A_{sym}, LS	0.7		25	
	0.52	1, II, A_{rec}, LS	0.7		32	
Chiquita Canyon C	0.09	1, I, A_{sym}, LS	0.4		10	
	0.18	1, II, A_{sym}, LS	0.4		14	
	0.21	1, II, A_{rec}, LS	0.4		16	
Chiquita Canyon D	0.10	1, I, A_{rec}, LS	0.4	44	8	
	0.10	1, I, A_{rec}, LS	0.4	50	4	
	0.23	1, II, A_{rec}, LS	0.4	75	8	
	0.23	1, II, A_{rec}, LS	0.4	80	4	
	0.24	2, I, A_{sym}, LS	0.4	76	8	
	0.24	2, I, A_{sym}, LS	0.4	81	4	
	0.25	2, I, A_{rec}, LS	0.4	78	8	
	0.25	2, I, A_{rec}, LS	0.4	83	4	
Chiquita Canyon C	0.18	1, II, A_{sym}, WS+LS	0.4		8	34
	0.18	1, II, A_{sym}, WS+LS	0.4		4	42
	0.21	1, II, A_{rec}, WS+LS	0.4		8	38
	0.21	1, II, A_{rec}, WS+LS	0.4		4	44
Lopez Canyon C Cross Section A - A'	0.23	1, I, A_{sym}, WS+LS	0.7		25	35
	0.23	1, I, A_{sym}, WS+LS	0.7		20	41
	0.25	1, I, A_{rec}, WS+LS	0.7		25	37
	0.25	1, I, A_{rec}, WS+LS	0.7		20	43
	0.36	1, II, A_{sym}, WS+LS	0.7		25	49
	0.36	1, II, A_{sym}, WS+LS	0.7		20	54

ANALYSIS ASSUMPTIONS:
1	-	one dimensional dynamic analysis
2	-	two dimensional dynamic analysis
I	-	average modulus reduction and damping curve (Kavazanjian and Matasovic 1995)
II	-	upper bound modulus reduction and lower bound damping curve (PI = 200 clay, Vucetic and Dobry 1991)
A_{rec}	-	modified closest recorded rock motion
A_{sym}	-	synthetic rock motion
WS	-	slip surface passing entirely through waste fill
LS	-	slip surface along liner interfaces
WS+LS	-	slip surface partially along liner interface and partially through waste fill

Another series of analyses was performed for the geosynthetic-lined landfills assuming a failure plane passing entirely along the geosynthetic liner surface to estimate dynamic geosynthetic interface strengths. Table 3 shows dynamic interface friction angles back-calculated at the Lopez Area C (Cross sections A-A' and B-B') and Chiquita Canyon C landfills. Lopez Canyon Area C has a textured geosynthetic interface along the base and smooth geosynthetic interface along side slopes. For the Lopez Canyon cross sections, the dynamic interface friction angle for the base liner required to maintain a safety factor of one ranged from 14° to 32°, assuming the dynamic interface friction angle of the side slope was 8°. In fact, for seismic coefficients less than 0.36 g, use of the liner shear strength parameters commonly employed in practice ($\phi = 8^{\circ}$ smooth HDPE interface and $\phi = 25^{\circ}$ textured HDPE interface), results in a safety factor greater than or equal to one for these slopes. Thus, these slopes would be expected to be stable during the earthquake, which is consistent with the observed performance during the earthquake. The Chiquita Canyon C landfill is lined with a smooth geosynthetic interface along the base and side slopes, and dynamic friction angles required to maintain a calibrated factor of safety of one using the observed deformation factor at this landfill ranged from 10° to 16°. The back-calculated friction angles are based on the assumption that some movement occurred at this landfill during the earthquake, which is consistent with the damage observed at this landfill.

Table 3 also shows the results of the back-analysis of the stability of the Chiquita Canyon D landfill. This landfill has a clay liner along the base and a geosynthetic liner along the landfill slopes. These analyses show that the stability of the Chiquita Canyon D landfill is more sensitive to the strength of the clay liner than the strength along the geomembrane interface. This is not unreasonable, as a majority of the critical slip surface lies along the clay liner interface. EMCON Associates (1994) estimated the shear strength parameters for the clay liner to be c = 62 kPa and $\phi = 0$. For reasonable values of the seismic coefficient (> 0.10), Table 3 indicates that this slope should have experienced some permanent deformations, which is consistent with the observed deformations and damage to the liner system at Canyon D. The results of these back-analyses generally support reported values of the liner interface shear strengths and suggest that these values, which were measured in the laboratory, appear to be reasonable representations of the shear strength of liner interfaces during seismic loading.

For two geosynthetically lined waste units, the critical dynamic slip surface passed partially along the liner surface and partially through the waste fill. Stability along these potential sliding masses is affected by the shear strength of the liner as well as the shear strength of the waste fill. At the Chiquita Canyon C landfill, the critical failure surface was along the side slope liner system passing through the waste fill approximately 150 m downslope. Assuming that the dynamic strength of the smooth geosynthetic liner interface is 8°, the dynamic strength of the waste fill ranges from 34° to 38°, which is consistent with the estimates of the dynamic friction angle for waste fill determined at the unlined landfills (Table 3). At Lopez Canyon Disposal Area C (Section A-A'), the critical failure plane was located near the waste slope's free face. Assuming that the static and dynamic strength of the textured geosynthetic is equal ($25°$), the dynamic strength of the waste fill ranged from $35°$ to $49°$. The larger dynamic waste fill friction angle was calculated from an analysis utilizing a higher estimate of the seismic coefficient, and this value of the friction angle may not be conservative. The results shown in Table 3 indicate that the dynamic friction angle of the waste fill is sensitive to the value of the dynamic interface friction angle along the liner interfaces.

CONCLUSIONS

The general performance of solid waste landfills during the Northridge Earthquake was encouraging. None of the solid waste landfills surveyed showed any signs of major damage. However, a geosynthetic-lined landfill close to the zone of energy release experienced significant damage evidenced by tears in the geomembrane liner. The geosynthetic liner systems at two other landfills at similar distances from the zone of energy release were undamaged. However, these landfills suffered moderate damage evidenced by cracking in interim cover soil at waste/natural ground transitions, breaking of gas extraction system header lines, and a loss of power to the gas collection system. Several unlined and soil lined landfills also experienced moderate damage, as evidenced by cracking in interim cover soils and damage to the gas collection systems. These observations are in general agreement with observations of landfill performance after the 1987 Whittier Narrows and 1989 Loma Prieta earthquakes.

Most landfills within 30 km of the zone of energy release experienced some form of cracking in the soil cover. Beyond approximately 40 km from the zone of energy release, little to no damage was observed. Cracking in cover soils may have resulted from one or more of the following: (a) brittle cracking of the stiffer soil veneer overlying the more ductile waste fill, (b) cracking resulting from the difference in relative stiffness between the softer waste fill and the stiffer adjacent natural ground, (c) seismically induced settlement of the waste fill, (d) limited downslope movement, and (e) cracking caused by the build up of gas underneath the soil cover due to the rapid release of gas produced by the shaking and/or temporary loss of the gas extraction system. Brittle cracking was observed at many of the surveyed landfills, especially at or near a free face or near changes in geometry, where an accumulation of transient, seismically induced strains would be expected. For unlined landfills, it is difficult to differentiate cracking associated with ground shaking induced settlements or incoherent motions from cracking at the back of the waste fill due to limited downslope movement along a failure zone. Observations available to date indicate that it is unlikely that any of the observed cracks resulted from a build-up of landfill gas after the Northridge Earthquake. However, the temporary loss of a waste landfill's gas extraction system is an important consideration because of the potential for fire or explosion.

An extensive set of static and dynamic stability analyses were performed to attempt to characterize the dynamic shear strength of solid waste and geosynthetic liner interface materials. Due to the uncertainty with the input rock motions, waste fill properties and analytical assumptions, these strength estimates should be regarded as preliminary. For a safety factor of 1.2, conservative static friction angles of waste fill ranged from $19°$ to $35°$. At unlined landfills, the results of pseudostatic stability analyses indicated that the dynamic strength of waste fill was higher than the conservative estimate of static strength. The solid waste dynamic friction angle estimated from this study ranged from $30°$ to $40°$. Pseudostatic stability analyses of geosynthetic-lined landfills resulted in back-calculated values for the dynamic friction angle of between $10°$ to $16°$ for smooth HDPE geomembrane interfaces and $14°$ to $32°$ for textured HDPE geomembrane interfaces. These values are consistent with those reported in the literature from laboratory testing for similar interfaces and materials.

ACKNOWLEDGMENTS

Financial support was provided by the National Science Foundation under Grants CMS-9416261 and BCS-9157083 and by the David and Lucile Packard Foundation. This support is gratefully acknowledged. The authors would like to thank Robert Anderson, John Clinkenbeard and Darryl Petker of the California Integrated Waste Management Board, Pete Mundy of CDM Federal Programs and Charles Dowdell and John Hower of the Los Angeles County Sanitation Districts, all of whom visited several of the sites mentioned in this paper and generously provided their field reconnaissance data. Also, special thanks are extended to Rod Nelson of the California Regional Water Quality Control Board.

APPENDIX I

REFERENCES

Anderson, D.G., Hushmand, B., and Martin, G.R. (1992), "Seismic Response of Landfill Slopes", Stability and Performance of Slopes and Embankments - II, ASCE Geotechnical Special Publication No. 31, New York, NY, pp. 973-989.

Augello, A.J., Bray, J.D., Leonards, G.A., Repetto, P.C., and Byrne, R.J. (1995), "Response of Landfills to Seismic Loading", Proceedings of Geoenvironment 2000, ASCE Geotechnical Special Publication No. 46, New York, NY, pp. 1051-1065.

Augello, A.J., Bray, J.D., Matasovic, N., Kavazanjian, E., Jr., and Seed, R.B. (1995), "Solid Waste Landfill Performance During the 1994 Northridge Earthquake" Proceedings Third International Conference on Recent Advances in Geotechnical Earthquake Engineering, St. Louis, Mo, Vol. 3 pp. 163-169.

Buranek, D. and Prasad, S. (1991), "Sanitary Landfill Performance During the Loma Prieta Earthquake", Proceedings Second International Conference on Recent Advances in Geotechnical Earthquake Engineering and Soil Dynamics, St. Louis, Mo, pp. 1655-1660.

California EPA (1994), "Observations of Landfill Performance, Northridge Earthquake of January 17, 1994", California Integrated Waste Management Board, Closure and Remediation Branch, Sacramento, CA.

CDM Federal Programs Corp. (1995), "Geotechnical Monitoring Data Compilation and Interpretation Report for December 1987 to October 1994, Operating Industries, Inc Landfill", Report to the U.S. Environmental Protection Agency, February.

Chang, S., Bray, J.D., and Seed, R.B. (1996), "Engineering Implications of Ground Motions from Northridge Earthquake", Bulletin of the Seismological Society of America, Northridge Special Issue, February, In Press.

CH2M Hill (1988), "Literature Review and Existing Data Review Technical Memorandum Geotechnical Task RI/FS2, South Parcel, Operating Industries, Inc., Monterey Park, CA." Report to the U.S. Environmental Protection Agency, February.

Coduto, D.P. and Huitric, R. (1990), "Monitoring Landfill Movements using Precise Instruments", In: Geotechnics of Waste Fills - Theory and Practice, ASTM STP 1070, A. Landva and G.D. Knowels Editors, American Society for Testing Materials, Philadelphia, PA, pp. 358-370.

Darragh, R., Cao, T., Cramer, C., Graizer, V., Huang, M., and Shakal, A. (1994 and 1995), "Processed CSMIP Strong-Motion Records from the Northridge, California Earthquake of January 17, 1994: Releases 1 through 11", California Division of Mines and Geology, Sacramento, CA.

Derian, L., Gharios, K.M., Kavazanjian, E., Jr., and Snow, M.S. (1993), "Geosynthetics Conquer the Landfill Law", Civil Engineering, ASCE, December.

Earth Technology Corporation (1988), "In-Place Stability of Landfill Slopes, Puente Hills Landfill, Los Angeles, California", Report No. 88-614-1, The Earth Technology Corporation, Long Beach, CA.

Earthquake Engineering Research Institute (1995), "The Northridge Earthquake Reconnaissance Report, Vol. 1", J.F. Hall Editor, Earthquake Spectra, Vol. 11, Supplement C, April.

EMCON Associates (1990), "Design Report for Construction of Chiquita Canyon C Landfill, Cells 1 and 2, Lining and Leachate Collection and Removal Systems", January.

EMCON Associates (1991), "Final Closure and Post-Closure Maintenance Plans Primary and Canyon B Landfills, Chiquita Canyon Landfill, Los Angeles County, CA", June, 1991, revised April, 1992.

EMCON Associates (1994), "Northridge Earthquake Seismic Evaluation, Chiquita Canyon Landfill", April.

GeoSyntec (1994), "Assessment of Performance of the Disposal Area C Liner System during the Northridge Earthquake of 17 January 1994 and of Damage to Geotextile and Geonet observed on 18 January 1994, Lopez Canyon Sanitary Landfill, Lake View Terrace, Los Angeles", Memorandum to Solid Waste Management Division, Bureau of Sanitation, City of Los Angeles, GeoSyntec Consultants, Hunting Beach, CA, 6 p.

GeoSyntec (1995), "Stability of Existing Landfill Refuse Slopes Subject to Surcharge Loading, Sunshine Canyon Sanitary Landfill, Sylmar California", GeoSyntec Consultants, Hunting Beach, CA, 10 p. plus Appendices.

Harding Lawson Associates (1986), "Disposal Site Information, Chiquita Canyon Landfill, Los Angeles County, CA", June.

Hudson, M., Idriss, I.M., and Beikae, M. (1994), "User's Manual for QUAD4M", Center for Geotechnical Modeling, Department of Civil and Environmental Engineering, University of California at Davis, Davis, CA.

Idriss, I.M. (1991), "Procedures for Selecting Earthquake Ground Motions at Rock Sites", A Report to the National Institute of Standards and Technology, University of California at Davis, September, revised March 1993.

Idriss, I.M. and Sun, J.I. (1992), "User's Manual for SHAKE91", Center for Geotechnical Modeling, Department of Civil and Environmental Engineering, University of California at Davis, Davis, CA.

IT Corporation (1994), "Final Closure and Post-Closure Maintenance Plan for Toyon Canyon Sanitary Landfill", October.

Johnson, M.E., Lundy, J., Lew, M., and Ray, M.E. (1991), "Investigation of Sanitary Landfill Slope Performance During Strong Ground Motion from the Loma Prieta Earthquake of October 17, 1989", Proceedings Second International Conference on Recent Advances in Geotechnical Earthquake Engineering and Soil Dynamics, St. Louis, Mo, April, pp. 1701-1706.

Kavazanjian, E., Jr., Hushmand, B. and Martin, G. R. (1991), "Frictional Base Isolation using a Layered Soil-Synthetic Liner System", Proceedings Third U.S. Conference on Lifeline Earthquake Engineering, ASCE, Los Angeles, CA, pp. 1140-1151.

Kavazanjian, E., Jr. and Matasovic, N. (1995), "Seismic Analysis of Solid Waste Landfills", Proceedings of Geoenvironment 2000, ASCE Geotechnical Special Publication No. 46, New York, NY, pp. 1066-1080.

Kavazanjian, E., Jr., Matasovic, N., Bonaparte, R., and Schmertmann, G.R. (1995), "Evaluation of MSW Properties for Seismic Analysis", Proceedings of Geoenvironment 2000, ASCE Geotechnical Special Publication No. 46, New York, NY, pp. 1126-1141.

Lindvall Richter Benuska (1994), "Processed LADWP Power System Strong-Motion Records from the Northridge, California Earthquake of January 17, 1994", Draft Report, Prepared for the Los Angeles Department of Water and Power, LRB NO. 007-0027.

Makdisi, F.I. and Seed, H.B. (1978), "Simplified Procedure for Estimating Dam and Embankment Earthquake-Induced Deformations", Journal of the Geotechnical Engineering Division, ASCE, Vol. 104, No. GT7, pp. 849-867.

Matasovic, N. (1993), "Seismic Response of Composite Horizontally-Layered Soil Deposits", P.h.D. Dissertation, Department of Civil and Environmental Engineering, University of California at Los Angeles, 483 pp.

Matasovic, N., Kavazanjian, E., Jr., Augello, A.J., Bray, J.D., and Seed, R.B. (1995), "Solid Waste Landfill Damage Caused by 17 January 1994 Northridge Earthquake", In: Woods, M.C. and Seiple, W.R., eds., The Northridge California Earthquake of 17 January 1994, California Department of Conservation, Division of Mines and Geology Special Publication 116, pp. 61-69.

Mitchell, J.K., Seed, R.B. and Seed, H.B. (1990), "Kettleman Hills Waste Landfill Slope Failure I: Liner System Properties", Journal of the Geotechnical Engineering Division, ASCE, Vol. 116, No. 4, pp. 647-668.

Mundy, P.K., Bastani, S.A., Brick, W.D., Clark, J., Nyznyk, J.P., and Herzig, R. (1995), "Results of Geotechnical Monitoring of an Instrumented Landfill in Southern California", ASCE Specialty Conference on Landfill Closures, ASCE National Convention, San Diego, CA, October, In Press.

Orr, W.R. and Finch, M.O. (1990), "Solid Waste Landfill Performance During the Loma Prieta Earthquake", In: Geotechnics of Waste Fills - Theory and Practice, ASTM STP 1070, A. Landva and G.D. Knowels Editors, American Society for Testing Materials, Philadelphia, PA, pp. 22-30.

Pacific Engineering and Analysis (1995), "Processing and Spectral Analysis of Eighteen Earthquakes Recorded at Operating Industries, Inc., Monterey Park, CA." Report to the U.S. Environmental Protection Agency, March.

Pacific Engineering and Analysis (1995), Synthetic input rock motions for the Northridge Earthquake, generated by Dr. Walt Silva.

Schneider, J.F., Silva, W.J., and Stark, C. (1993), "Ground Motion Model for the 1989 M 6.9 Loma Prieta Earthquake Including Effects of Source, Path, and Site", Earthquake Spectra, Earthquake Engineering Research Institute, Vol. 9, No. 2.

Seed, R.B. and Boulanger, R. (1992), "Smooth HDPE-Clay Liner Interface Shear Strength: Compaction Effects", Journal of the Geotechnical Engineering Division, ASCE, Vol. 117, No. 4, pp. 686-693.

Sharma, H.D. and Goyal, H.K. (1991), "Performance of a Hazardous Waste and Sanitary Landfill Subjected to Loma Prieta Earthquake", Proceedings Second International Conference on Recent Advances in Geotechnical Earthquake Engineering and Soil Dynamics, St. Louis, Mo, April, pp. 1717-1725.

Silva, W.J. (1993), "Factors Controlling Strong Ground Motions and their Associated Uncertainties", ASCE Symposium on High Level Nuclear Waste Repositories, pp. 132-161.

Stewart, J.P., Bray, J.D., Seed, R.B., and Sitar, N. (1994), "Preliminary Report on the Principal Geotechnical Aspects of the January 17, 1994 Northridge Earthquake", Report No. UCB/EERC - 94/08, Earthquake Engineering Research Center, University of California at Berkeley, Berkeley, CA, June.

U.S. Geological Survey (1994), "USGS National Strong-Motion Program Digitized Records from the Northridge, California Earthquake of January 17, 1994", from Internet address agram.wr.usgs.gov.

Vucetic, M. and Dobry, R. (1991), "Effect of Soil Plasticity on Cyclic Response", Journal of the Geotechnical Engineering Division, ASCE, Vol. 117, No. 1., pp. 89-107.

Waste Management of North America, Inc. (1991), "Preliminary Closure Plan and Preliminary Postclosure Maintenance Plan, Bradley Avenue East and West Sanitary Landfill, Sun Valley, California".

Woodward Clyde Consultants (1988), "Inspection of the OII Site following the 1 October 1987 Whittier Narrows Earthquake", report prepared for Camp Dresser McKee, Woodward Clyde Consultants, Santa Ana, CA, February.

Wright, S.G. (1990), *UTEXAS3 A Computer Program for Slope Stability Calculations*, Austin, Texas.

Yegian, M.K. and Lahlaf, A.M. (1992), "Dynamic Interface Shear Strength Properties of Geomembranes and Geotextiles", Journal of the Geotechnical Engineering Division, ASCE, Vol. 118, No. 5, pp. 760-779.

Earthquake Ground Motions for Landfill Design

Kenneth W. Campbell, M. ASCE[1]

Abstract

Municipal Solid Waste Landfill Facilities have unique earthquake response characteristics that should be considered when developing earthquake ground motions for seismic design. Because of the need to quantify and limit damage during strong ground shaking, it is important to define design ground motions in terms of their amplitude, frequency content, energy content, and duration. In this paper, the important factors that should be considered when developing design ground motions for landfill design are described and demonstrated with preliminary results from an ongoing seismic design study of the OII Landfill Superfund site in Monterey Park, California.

Introduction

Minimum Federal design standards for Municipal Solid Waste Landfill Facilities (MSWLF), which took effect October 9, 1993, are contained in Subtitle D, Section 258.14 of Title 40 of the Code of Federal Regulations (40 CFR 258.14). These regulations require that containment systems for landfills constructed in *seismic impact zones* be designed to resist the maximum horizontal acceleration (MHA) in "lithified earth material" (i.e., rock) at the site. Seismic impact zones are defined by the Environmental Protection Agency (EPA) as areas with a 10% or greater probability that the MHA in lithified earth material will exceed 0.1g in 250 years. This probability corresponds to an average return period of approximately 2,500 years. By this definition, approximately half of the United States is contained within an EPA seismic impact zone.

The MHA can be determined from existing seismic hazard maps, such as those published by the U.S. Geological Survey in 1982 and 1990 (Algermissen and others, 1982, 1990), or from a site-specific probabilistic seismic hazard analysis.

---

[1] Associate and Senior Technical Manager, EQE International, Inc., 2942 Evergreen Parkway, Suite 302, Evergreen, Colorado, 80439

The latter analysis requires an assessment of the seismotectonics, seismicity, and attenuation characteristics of the region surrounding the site. Procedures for conducting such an analysis are summarized by the EERI Seismic Risk Committee (1989). If the MSWLF is located on unconsolidated sediments, a dynamic site-response analysis is required to estimate the ground motions at the base of the landfill. Whether or not a site-response analysis is necessary, time histories are required in order to determine design ground motions within and on the landfill. Since the MHA is defined probabilistically, there is no specific earthquake magnitude, source-to-site distance, or duration that corresponds to this motion. This is particularly true if the MHA is estimated from an existing seismic hazard map. Thus, the development of appropriate time histories to use for design is not a trivial matter.

Time histories can be developed synthetically—so-called artificial accelerograms—or selected and scaled from a catalog of available digital accelerograms. In either case, parameters other than the MHA, such as magnitude and source-to-site distance, are required in order to estimate the appropriate duration and frequency content of the ground motions. If no other information is available, these parameters can be conservatively derived from the magnitude of the maximum credible earthquake (MCE) and the distance to the fault or seismotectonic province that produces the highest ground motions at the site in the frequency range of interest from a comparison of deterministically estimated response spectra, For this purpose, both peak and energy spectra should be evaluated in order to include the effects of duration in the selection process. This approach will be extremely conservative in areas of low seismicity in which the 2,500-year MHA is controlled by earthquakes significantly smaller than the MCE. In such cases, it is desirable to disaggregate the results to determine the mean magnitude and mean distance and/or a suite of controlling magnitudes and distances that adequately represent the hazard, then use these parameters to develop appropriate time histories. This procedure, however, requires that a site-specific seismic hazard analysis be performed, thus precluding the use of a published seismic hazard map to estimate the MHA

A case study of the Operating Industries, Inc. (OII) Landfill Superfund site in Monterey Park, California, is used to demonstrate the development of design earthquake ground motions using the EPA recommended procedures. Because of its classification as a Superfund site, the OII landfill is not subject to regulations under Subtitle D. Instead, it is being designed to deterministic criteria recommended by EPA and its Technical Review Panel. However, for demonstration purposes, probabilistic procedures consistent with the current EPA requirements will be compared to the deterministic procedures used in the existing preliminary design of the landfill containment systems to contrast these two design methodologies.

OII Landfill

The OII Landfill is located in Monterey Park, California, within the margin uplands of the Los Angeles Basin, a coastal plain approximately 80 kilometers long by 30 kilometers wide located in southern California. The basin is bounded on three sides by various hills and mountains and has a mild westerly seaward gradient. The major features in the local vicinity of the OII Landfill are the Montebello and Monterey Hills. These hills consist of a chain of uplands that were formed as a result of north-south compressional forces associated with the regional stress field. The hills are surface expressions of the Elysian Park Fold and Thrust Belt, a recently postulated regional deep-seated structural mechanism that includes the Elysian Park blind thrust fault, source of the 1987 Whittier Narrows earthquake (M_W = 6.0). The earthquake occurred only a few kilometers northeast of the OII Landfill causing only minor cracking, even though peak accelerations as high as 0.48g were recorded in the vicinity of the site on sediments similar to those beneath the trash prism.

The major exposed bedrock units in the local site area include the Miocene age marine Puente and Topanga Formations, the late Pliocene age marine Pico Time Unit, and the undifferentiated Pleistocene age fluvial Lakewood and San Pedro Formations. Recent alluvium and other surficial deposits (i.e., soil and colluvium) exist locally throughout the region; however, such deposits are not present at the OII site. The base of the landfill is founded principally on the Pliocene age Pico Time Unit and to a lessor extent on the Pleistocene age Lakewood and San Pedro Formations. These bedrock units have a shear-wave velocity ranging from about 365 to 750 m/s and are, therefore, classified as a soft rock (e.g., Borcherdt, 1994).

Seismic Design Methodology for the OII Landfill

Because of its designation as a Superfund site, the OII Landfill is not subject to design under Subtitle D of 40 CFR 258.14. Prior to October 1993, MSWLFs were typically designed to median ground motions associated with the controlling Maximum Credible Earthquake (MCE) on known faults in the vicinity of the site. However, because of the designation of the OII Landfill as a Superfund site and because of its location in a populated metropolitan area, the EPA and its Technical Review Panel recommended that it be designed to 84th-percentile (i.e., the median plus one standard deviation) ground motions from the controlling MCE. They also recommended that ground-shaking duration and distance to the site be explicitly considered when selecting the design earthquakes because of the potentially unique response characteristics of the trash prism.

Although the final design ground motions had not been approved as of the date of this paper, consistent with EPA's initial recommendations, the methodology used to develop the preliminary design ground motions consisted of four major steps: (1) identifying all potentially controlling faults in the region within 100 kilometers of the OII site, (2) estimating peak-amplitude and kinetic-energy ground acceleration and elastic response spectra at the site for each of the potentially controlling faults, (3) identifying the controlling far-field, intermediate-field, and near-field earthquakes that contribute the highest spectral amplitudes or kinetic energy at the site in a broad range of frequencies, and (4) developing design response spectra and design accelerograms for each of the controlling earthquakes. For this purpose, far field, intermediate field, and near field were defined as distances > 50 km, distances > 20 km and ≤ 50 km, and distances ≤ 20 km from the site, respectively.

Potentially Controlling Faults at OII

Potentially controlling faults in the far-field, intermediate-field, and near-field regions were selected from among the regional faults by choosing the closest faults with the largest MCE magnitudes in each region. This selection was done independently for strike-slip and reverse/reverse-oblique faults, because some of the attenuation relationships used in the study predict that these two types of earthquakes will have different amplitudes and attenuation characteristics. A fault was eliminated only if there was a similar type of fault closer to the site within the same region with an equal or larger MCE. The screening process resulted in the selection of ten potentially controlling faults: two in the far-field region, three in the intermediate-field region, and five in the near-field region. These faults are listed in Table 1.

Preliminary Design Earthquakes at OII

The peak horizontal ground acceleration (A_P) and the peak horizontal spectral acceleration (S_P) for each of the potentially controlling earthquakes were estimated from the MCE magnitude and source-to-site distance using three representative attenuation relationships (Boore and others, 1993; Sadigh and others, 1993; Campbell and Bozorgnia, 1994). The relationship of Campbell and Bozorgnia, which was developed only for peak horizontal ground acceleration, was supplemented with attenuation relationships for normalized response spectra developed by Campbell (1993) using a similar data base and methodology to that used by Campbell and Bozorgnia. For purposes of the present study, preliminary design ground motions were defined in terms of the geometric mean (i.e., the mean of the logarithms) of the estimates from the three relationships.

Since each attenuation relationship uses a different definition of source-to-site distance, three different distance measures were estimated for each potentially controlling fault. For the relationship of Boore and others (1993), distance was measured from the surface projection of the assumed fault rupture. For the relationship of Sadigh and others (1993), distance was measured from either the fault trace, if the earthquake was assumed to rupture to the surface (i.e., its magnitude was greater than about 6¼) and this trace was the closest part of the rupture to the site; or to the rupture surface at depth, if the earthquake was not expected to rupture to the surface or this was the closest part of the rupture to the site. For the relationships of Campbell (1993) and Campbell and Bozorgnia (1994), distance was measured from the assumed seismogenic rupture zone on the fault, which for all but the Elysian Park thrust fault was conservatively assumed to occur at a depth of 3 km, as recommended by the authors. For the Elysian Park thrust, this depth was taken to be 10 km, the shortest distance to the fault plane at depth.

The site classifications used with each of the attenuation relationships were chosen to be as consistent as possible with the geologic description, age, and shear-wave velocity of the bedrock beneath the OII site. Accordingly, estimates were made for sites described as *Site Class B* by Boore and others (1993), as *Rock* by Sadigh and others (1993), and as *Soft Rock* by Campbell (1993) and Campbell and Bozorgnia (1994). Site Class B of Boore and others represents sites having an average shear-wave velocity of 360 to 750 m/sec in the upper 30 m of the deposit. The shear-wave velocity of the bedrock below the OII site falls within this range. The spectral relationships of Campbell (1993) require an estimate of the depth to basement rock beneath the site. This depth was estimated to be about 2.5 km from a basement-rock contour map for the Los Angeles Basin developed by Yerkes and others (1965). Estimates of A_P and S_P for oscillator periods of 0.3, 1 and 2 sec for each of the potentially controlling faults are given in Table 1.

The kinetic energy associated with the ground acceleration and the 5%-damped acceleration response spectra for each of the controlling faults were estimated from root-mean-square (RMS) amplitude and duration from the expression,

$$Y_{KE} = \tfrac{1}{2} Y_{RMS}^2 \, T_{RMS}$$

where Y_{KE} is the kinetic energy per unit mass of the ground motion (A_{KE}) or response spectrum (S_{KE}), Y_{RMS} is the RMS value of Y, T_{RMS} is the duration of Y_{RMS}, and RMS amplitude is related to peak amplitude by the expression,

$$Y_{RMS} = Y_P/F_P$$

where F_P is the peak factor.

The RMS duration was estimated from seismological theory and adjusted for the duration of a single-degree-of-freedom system based on relationships proposed by Boore (1983), Boore and Joyner (1984), and Herrmann (1985). This duration is dependent on both magnitude and distance, being longer for larger magnitudes and longer distances. The calculated RMS duration represents the rupture duration of a simple circular source modified for wave-propagation effects. It accounts for the increase in duration with distance resulting from surface waves and supercritically reflected waves using a simple crustal model (Herrmann, 1985). Since the calculated duration does not take into account the increased duration that can result from multiple sources (i.e., an earthquake composed of more than one subevent) or from scattering, multiple reflections, or basin response effects that can be expected for more realistic crustal models, it should be considered a minimum estimate. However, because only relative amplitudes of kinetic energy were required for determining the controlling earthquakes, this minimum estimate was considered adequate.

The peak factor was estimated from RMS duration and the number of zero crossings using random vibration theory. The RMS duration and the peak factor were estimated from the MCE magnitude, the minimum source-to-site distance of each potentially controlling fault, and the velocity structure beneath the site using the computer code RASCAL developed by Silva and Lee (1987). The velocity structure was estimated by fitting a power-law function to the depth to the upper three layers of the velocity model developed by Hauksson and Jones (1989) for the Whittier Narrows area. The amplification spectrum associated with this velocity structure was approximated by the methodology suggested by Joyner and others (1981). Estimates of A_{KE} and S_{KE} for oscillator periods of 0.3, 1 and 2 sec for each of the potentially controlling faults are given in Table 1.

A comparison of the 84th-percentile estimates of A_P, A_{KE}, S_P, and S_{KE} for the two potentially controlling far-field faults indicates that in all cases the San Andreas fault produces the largest ground motions (Table 1). Therefore, the MCE on this fault was selected as the preliminary design earthquake for the far-field region. A similar comparison for the three potentially controlling intermediate-field faults indicates that in all cases the Newport-Inglewood fault produces the highest peak ground motions. However, the Elsinore fault generates almost the same amount of kinetic energy as the Newport-Inglewood fault at long periods. Therefore, the MCEs on both of these faults were selected as alternative candidates for the preliminary design earthquake in the intermediate-field region. Finally, a comparison of the 84th-percentile estimates of A_P, A_{KE}, S_P, and S_{KE} for the five potentially controlling near-field faults indicates that in all cases the Whittier fault produces the largest ground motions (Table 1). Therefore, the MCE on this fault was selected as the preliminary design earthquake in the near-field region.

The largest estimated values of the median and 84th-percentile maximum horizontal acceleration (MHA) on bedrock at the OII site are $0.48g$ and $0.75g$ for the Whittier fault ($M_W = 7$, Distance = 5 km). In contrast, estimates of the 2,500-year MHA based on the seismic hazard maps of Algermissen and others (1982, 1990) are approximately $0.8g$ and greater. Thus, the 2,500-year estimate of MHA at the OII site is significantly greater (> 67%) than the median estimate of MHA from the controlling MCE, the standard by which MSWLFs were designed prior to October 1993. In fact, the 2,500-year estimate of MHA is even greater than the 84th-percentile estimate of the MHA for the controlling MCE, the standard by which critical structures such as dams and nuclear power plants are designed.

Preliminary Design Accelerograms at OII

Accelerograms were developed for each of the preliminary design earthquakes by scaling existing corrected time histories to approximate the preliminary design spectra for each event. Time histories were selected to represent earthquakes having similar magnitudes and distances as these events. This assured that the accelerograms had approximately the same amplitude, frequency content, energy content, and duration as that of the preliminary design earthquake. Due to the limited number of processed accelerograms, it was not always possible to select time histories recorded on site conditions similar to the OII site. Instead, recordings were selected to have similar spectral shapes as those predicted from the attenuation relationships, independent of site conditions.

The time histories were scaled to have the same peak ground acceleration as that predicted for the preliminary design event in each region. The response spectra from these scaled accelerograms were then compared to the mean predicted response spectra for each of the preliminary design earthquakes to insure reasonable agreement over a broad range of periods. This comparison indicated that the scaled accelerograms had response spectra that were in reasonable agreement with the mean predicted response spectra for oscillator periods of 0.2 to 2 sec. A list of the selected time histories along with their associated magnitudes and distances, strong-motion characteristics, and scaling factors are summarized in Table 2.

References

Algermissen, S.T., D.M. Perkins, P.C. Thenhaus, S.L. Hanson, and B.L. Bender (1982). "Probabilistic Estimates of Maximum Acceleration and Velocity in Rock in the Contiguous United States," *U.S. Geological Survey Open File Report 82-1033.*

Algermissen, S.T., D.M. Perkins, P.C. Thenhaus, S.L. Hanson, and B.L. Bender (1990). "Probabilistic Earthquake Acceleration and Velocity for the United States and Puerto Rico," *U.S. Geological Survey Miscellaneous Field Studies Map MF-2120.*

Boore, D.M. (1983). "Stochastic Simulation of High-Frequency Ground Motions Based on Seismological Models of the Radiated Spectra," *Bulletin of the Seismological Society of America*, Vol. 73, p. 1865-1894.

Boore, D. M., and W. B. Joyner (1984). "A Note on the Use of Random Vibration Theory to Predict Peak Amplitudes of Transient Signals," *Bulletin of the Seismological Society of America*, Vol. 74, p. 2035-2039.

Borcherdt, R.D. (1994). "Estimates of Site-Dependent Response Spectra for Design (Methodology and Justification)," *Earthquake Spectra*, Vol. 10, p. 617-653.

Boore, D. M., W. B. Joyner, and T. E. Fumal (1993). "Estimation of Response Spectra and Peak Accelerations From Western North American Earthquakes: An Interim Report," *U.S. Geological Survey Open-File Report 93-509.*

Campbell, K. W. (1993). "Empirical Prediction of Near-Source Ground Motion From Large Earthquakes," in *Proceedings, International Workshop on Earthquake Hazard and Large Dams in the Himalaya*, V.K. Gaur, ed., Jan. 15-16, New Delhi, Indian National Trust for Art and Cultural Heritage (INTACH), New Delhi, p. 93-103.

Campbell, K. W., and Y. Bozorgnia (1994). "Near-Source Attenuation of Peak Horizontal Acceleration From Worldwide Accelerograms Recorded From 1957 to 1993," in *Proceedings, Fifth U.S. National Conference on Earthquake Engineering*, July 10-14, 1994, Chicago, Earthquake Engineering Research Institute, Oakland, California, Vol. III, p. 283-292.

EERI Seismic Risk Committee (1989). "The Basics of Seismic Risk Analysis," *Earthquake Spectra*, Vol. 5, p. 675-702.

Hauksson, E., and L. M. Jones (1989). "The 1987 Whittier Narrows Earthquake Sequence in Los Angeles, Southern California: Seismological and Tectonic Analysis," *Journal of Geophysical Research*, Vol. 94, p. 9569-9589.

Herrmann, R. B. (1985). "An Extension of Random Vibration Theory Estimates of Strong Ground Motion to Large Distances," *Bulletin of the Seismological Society of America*, Vol. 75, p. 1447-1453.

Joyner, W. B., R. E. Warrick, and T. E. Fumal (1981). "The Effect of Quaternary Alluvium on Strong Ground Motion in the Coyote Lake, California, Earthquake of 1979," *Bulletin of the Seismological Society of America*, Vol. 71, p. 1333-1349.

Sadigh, K., C.-Y. Chang, N. A. Abrahamson, S. J. Chiou, and M. S. Power (1993). "Specification of Long-Period Ground Motions: Updated Attenuation Relationships for Rock Site Conditions and Adjustment Factors for Near-Fault Effects," in *Proceedings, Seminar on Seismic Isolation, Passive Energy Dissipation, and Active Control*, Applied Technology Council, Redwood City, California, Report No. ATC-17-1, Vol. 2, p. 59-70.

Silva, W. J., and K. Lee (1987). "WES RASCAL Code for Synthesizing Earthquake Ground Motions," in *State-of-the-Art for Assessing Earthquake Hazards in the United States*, U.S. Army Corps of Engineers Waterways Experiment Station, Vicksburg, Mississippi, Miscellaneous Paper S-73-1, Report No. 24.

Yerkes, R. F., T. H. McCulloh, J. E. Schoellhamer, and J. G. Vedder (1965). "Geology of the Los Angeles Basin California—An Introduction," *U.S. Geological Survey Professional Paper 420-A*.

TABLE 1

**Preliminary Estimates of the 84th-Percentile Horizontal Peak
and Kinetic Energy Ground Motions at the OII Landfill Site: Ground Acceleration
and 5%-Damped Acceleration Response Spectra at Periods of 0.3, 1 and 2 sec**

Fault	$R^{(1)}$ (km)	$MCE^{(1)}$ (M_W)	Horizontal Ground Motion $^{(2)}$							
			Peak Amplitude (g)				Kinetic Energy (g-cm)			
			A_P	$S_P(3)$	$S_P(1)$	$S_P(2)$	A_{KE}	$S_{KE}(.3)$	$S_{KE}(1)$	$S_{KE}(2)$
NEAR FIELD										
Alhambra Wash	2.1	6	0.64	1.25	0.48	0.18	0.045	0.253	0.064	0.011
Whittier	5.0	7	*0.75*	*1.58*	*0.80*	*0.35*	*0.154*	*0.806*	*0.300*	*0.091*
Coyote Pass	5.8	6	0.57	1.12	0.40	0.17	0.037	0.204	0.046	0.010
Raymond	9.0	6¾	0.57	1.19	0.55	0.25	0.073	0.383	0.123	0.040
Elysian Park Thrust	10.0	6¾	0.69	1.46	0.77	0.30	0.105	0.567	0.242	0.059
INTERMEDIATE FIELD										
Newport-Inglewood	22	7	*0.30*	*0.60*	*0.31*	*0.15*	*0.026*	*0.124*	*0.047*	*0.017*
Santa Monica	28	7	0.25	0.50	0.26	0.13	0.019	0.087	0.034	0.013
Elsinore	43	7½	0.20	0.40	0.25	0.13	0.020	0.082	*0.042*	*0.015*
FAR FIELD										
San Andreas (central)	52	8	*0.23*	*0.42*	*0.31*	*0.16*	*0.040*	*0.137*	*0.093*	*0.032*
Anacapa	53	7¼	0.15	0.29	0.17	0.09	0.009	0.036	0.017	0.007

(1) R=closest horizontal distance to the fault; MCE=maximum credible earthquake

(2) Values in bold italics identify the largest within the region (field)

TABLE 2

Characteristics of Preliminary Design Accelerograms for the OII Landfill Site

Fault	Earthquake/Year	Recording Station	Comp. (Az.)	Site Conditions	$M^{(1)}$	$R^{(1)}$ (km)	$T^{(2)}$ (sec)	$A_p^{(3)}$ (g)	Scale Factor
NEAR FIELD									
Whittier	Loma Prieta 1989	Santa Teresa Hills	315	Alluvium/ Serpentine	6.9 (7)	10 (5)	15	0.238 (0.75)	3.151
	Petrolia 1992	Rio Dell	272	Alluvium (15m)	7.1 (7)	15 (5)	20	0.386 (0.75)	1.945
INTERMEDIATE FIELD									
Newport-Inglewood	Loma Prieta 1989	Santa Teresa Hills	315	Alluvium/ Serpentine	6.9 (7)	10 (22)	15	0.238 (0.30)	1.261
Elsinore	Landers 1992	Amboy	360	Alluvium (3m)/Basalt	7.4 (7½)	64 (39)	40	0.115 (0.20)	1.739
FAR FIELD									
San Andreas	Landers 1992	Indio	090	Alluvium	7.4 (8)	40 (52)	60	0.109 (0.23)	2.101
	Synthetic	Caltech A1	—	—	8 (8)	50 (52)	70	0.204 (0.23)	1.130

(1) M=moment magnitude; R=closest horizontal distance to fault; Values in parentheses are preliminary design values

(2) Estimated duration of accelerogram excluding low-amplitude coda

(3) Recorded peak horizontal acceleration; Values in parentheses are preliminary 84th-percentile design values

Seismic Analysis of Solid Waste Landfills

Donald E. Del Nero, P.E.[1], Brian W. Corcoran, P.E.,[2] and Shobha K. Bhatia, Ph.D.[3]

Abstract

Specific provisions are included in Title 40, Part 258 (RCRA Subtitle D) of the 1991 United States (U.S.) Environmental Protection Agency (EPA) regulations governing municipal solid waste that dictate a new landfill unit or lateral expansion cannot be located in seismic impact zones unless the containment structure can withstand earthquake accelerations. Through the use of seismic response analyses on three existing landfill sites, two state-of-the-art landfills in New York State (NYS) and one existing landfill in Massachusetts, the authors noted that the selection of dynamic material properties and a representative earthquake motion are potentially the most important decisions a practitioner faces in seismic response analysis. Results of seismic response analyses by the authors and other researchers and practitioners are examined to evaluate input motion propagation with special emphasis on input motion frequency. Seismic response of the waste fill/foundation systems was assessed with "WESHAKE", which was adapted by the U.S. Army Corps of Engineers from the original "SHAKE" model (Schnabel et al., 1972). Multiple simulations were conducted with two U.S. west coast and two eastern North American earthquakes accelerograms. The authors found that peak bedrock accelerations alone, are inadequate to properly characterize seismic response. As exhibited by these simulations, the model predicted a damping of the high frequency eastern North American motions and an amplification of the low frequency western U.S. motions.

Introduction

As a result of the promulgation of Federal regulations requiring seismic analysis of landfills, the field of earthquake engineering has accelerated research to come up with standards for, and confidence in, analytical procedures. Bray et al. (1993) pointed out that seismic impact zones cover nearly one-half of the continental U.S., which will have a large impact on landfill analysis and design.

----

[1]Parsons Engineering Science, Inc., Liverpool, NY 13088
[2]O'Brien & Gere Engineers, Inc., Syracuse, NY 13221
[3]Assoc. Prof., Dept. of Civil & Envir. Engrg., Syracuse Univ., Syracuse, NY 13244

In NYS alone, these provisions are having a large impact. At typical liner system slopes of 33% and the inclusion of geosynthetics, even static conditions can be challenging for practitioners. In light of the environmental impacts from a liner system failure and a subsequent leachate excursion, it is incumbent on the geotechnical community to develop the expertise to ensure seismic response predictions are accurate. Towards that end, efforts such as those by the University of Southern California in their workshop on Research Priorities for Seismic Design of Solid Waste Landfills held in August, 1993 are to be commended.

The current challenges in solid waste landfill seismic analysis include a significant lack of knowledge and uncertainties in the following areas:

- Dynamic properties of waste and geosynthetic products;

- Base isolation and waste/liner/foundation interaction;

- Earthquake motion frequency content as it relates to seismic response;

- Landfill performance criteria;

- Case history performances; and

- Analytical procedures;

The area of earthquake motion frequency content is the subject of this paper. Nuttli (1981) and Atkinson (1987) point out that one of the primary differences anticipated between earthquakes in the eastern and western U.S. is the frequency content. Results are presented of research that consisted of the excitation of three existing landfill sites in the northeastern U.S. with two west coast earthquakes and two eastern North American earthquakes.

Site Descriptions

Three landfill sites located in the northeastern U.S. were evaluated for this study. Since NYS and the northeastern U.S. in general is one of the most seismically active areas east of the Rocky Mountains, it is important that practitioners consider seismic analysis in their landfill designs. The three sites analyzed are located in seismic impact zones and are subject to State and Federal seismic regulations governing solid waste landfill sites. Landfill cross-sections showing general site geology of the three sites are presented in Figure 1.

Site #1 is located in southeastern NYS and has a maximum probable horizontal acceleration of 0.12g as interpolated from map MF-2120 for a 250 yr. return period (Algermissen et al., 1991). This site consists of approximately 32 meters of municipal solid waste bearing on 19 meters of a predominantly dense to very dense sand overlying sandstone bedrock.

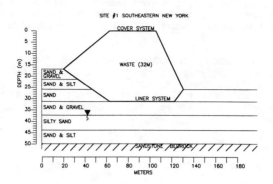

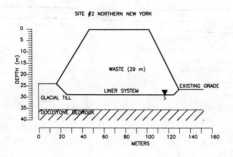

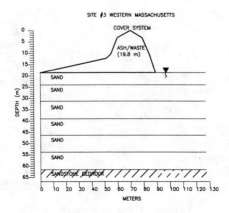

Figure 1. Landfill cross-sections for sites 1, 2 and 3.

Site #2 is located in the northern NYS and has a maximum probable horizontal acceleration of 0.21g as interpolated from map MF-2120 for a 250 yr. return period. The site consists of approximately 29 meters of municipal solid waste bearing on 7 meters of dense glacial till overlying dolostone bedrock.

Site #3 is located in western Massachusetts and has a maximum probable horizontal acceleration of 0.2g as interpolated from map MF-2120 for a 250 yr. return period. The site consists of approximately 20 meters of waste comprised of ash and ash mixed with bypass municipal solid waste. The waste mass bears on approximately 43 meters of loose to very dense sand overlying sandstone bedrock.

Model Input Parameters

Dynamic MSW/Soil Properties

The dynamic properties incorporated in the analyses are given on Figure 2. Additionally, modulus reduction and damping curves for the municipal solid waste (MSW) were taken from Singh and Murphy (1990) as recommended in the EPA (1995) guidelines. The approach used for developing the MSW density and shear wave velocities is consistent with that recommended by Kavazanjian et. al (1995). This consists of simulating the waste mass with a density and shear wave velocity gradient. The authors support this approach in light of the changes in MSW as it ages and decomposes under the weight of overlying waste.

The complicating factor in selecting site specific MSW parameters is that the MSW streams vary significantly across the U.S. Large disparities in recycling efforts across the Country have a direct impact on the short and long term properties of MSW. This must be a consideration in performing site specific seismic analyses.

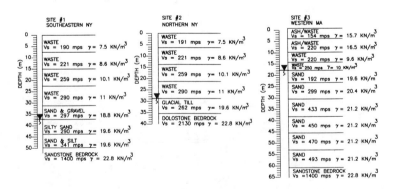

Figure 2. Site Profiles Analyzed.

But these correlations were not based on solid waste $(N_1)_6$. ?

SOLID WASTE LANDFILLS   65

The shear wave velocities for the landfill foundations were developed from correlations with standard penetration test blow counts. Sykora (1987) developed these correlations which include the depth at which the blow count was observed. The modulus reduction and damping curves selected for the analysis were obtained from the database included in the WESHAKE model literature.

Input Bedrock Motions

In light of the EPA (1995) guidance that recommends a suite of three records for a seismic evaluation and the east coast versus west coast motion comparative study, four earthquake motions were selected for the wave propagation analysis for the three sites. The motions selected are given on Figure 3. They were selected because of their inherent low and high frequency content, west coast versus east coast, respectively. Other factors (Bray et al., 1993) such as earthquake source mechanism, travel path geology, topographic effects, earthquake magnitude, distance from zone of energy release, and local soil conditions should also be considered in selecting the actual earthquakes to be used in a seismic analysis.

But not here!

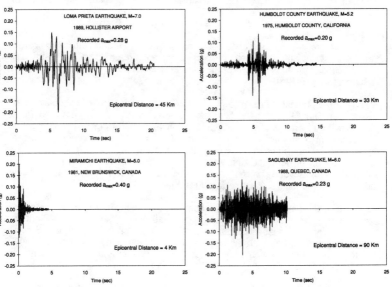

Figure 3. Earthquake acceleration time history for motions used in analyses. (All motions are scaled to peak acceleration of 0.2 g).

The motions from eastern North America seem to be enriched in high frequencies. This may be a result of the bedrock through which the seismic waves travel. Bray et al. (1993) points out that the characteristics of the rock through which seismic waves travel influence the frequency content of the seismic energy. Additionally, the bedrock in NYS is older, cooler, and more rigid than the bedrock in California (State of NY, 1991).

To facilitate comparisons between the different bedrock motions and site responses the four accelerograms were scaled to 0.2g. As pointed out, this is fairly consistent with the acceleration shown for these site localities on the seismic probability maps (Algermissen et al., 1991) often referenced for the U.S. for the 250 return period required by the EPA. An interesting observation (Jacob et al., 1994) is that most building codes and the current AASHTO bridge specifications use a 50 year return period, which is a less severe event.

> ? [10% chance of exceedance
> ∴ 50 yrs]

Results of One-Dimensional Analysis

One-dimensional analyses were performed for the three sites using the computer program WESHAKE, which estimates the seismic response of a layered soil deposit. The seismic responses of each site are characterized by estimates of peak acceleration in each layer of the soil/waste profile and response spectra for the cover system, liner system and bedrock.

The response of the different layers within the landfill soil/waste profiles in terms of maximum computed acceleration are presented in Table 1. The input bedrock motions applied to the sites were standardized to 0.2g to facilitate a comparative analysis.

An examination of Table 1 for the Loma Prieta, 1989 earthquake (west coast) for Site #1 shows the bedrock acceleration of 0.2g is amplified in all layers with an estimated peak acceleration at the landfill crest of 0.389g. Site #2 reacts in a similar way with an amplification of the bedrock motion to 0.348g at the landfill crest. The only difference is a slight deamplification from the glacial till through the first waste layer. The bedrock motion at Site #3 is amplified for several layers, then damped for several layers before being amplified at the landfill crest surface to 0.309g.

For the Saguenay, 1988 earthquake (east coast) Site #1 shows the scaled maximum bedrock acceleration of 0.2g generally being damped throughout the soil/waste profile to an acceleration at the landfill crest of 0.155g. Site #2 reacts in a similar way with the bedrock motion being damped to an acceleration at the landfill crest of 0.149g. Site #3 also shows a damping, with an acceleration at the crest of the landfill of 0.129g.

Considering all three site responses for the four earthquakes shows an interesting trend. When the sites are excited with the west coast motions (Loma Prieta, 1989 and Humboldt County, 1975) there is an amplification of the peak bedrock motions; whereas the east coast motions (Saguenay, 1988 and Miramichi, 1981) are damped as they propagate through the waste mass. These results provide an excellent comparison of the seismic response of a specific site to four distinctly different earthquake motions in terms of their frequency content.

Table 1. Maximum Acceleration Computed at Different Layers of Soil/Waste by WESHAKE

Depth, m	Material	Maximum Acceleration, g.			
		Loma Prieta	Humboldt County	Saguenay	Miramichi
Site #1 Southeastern New York					
0	Waste	0.389	0.317	0.155	0.11
7.62	Waste	0.267	0.231	0.082	0.071
15.2	Waste	0.228	0.218	0.103	0.093
22.8	Waste	0.224	0.216	0.119	0.114
31.7	Sand & Gravel	0.226	0.250	0.155	0.161
37.8	Silty Sand	0.215	0.181	0.119	0.135
44.5	Sand & Silt	0.214	0.182	0.142	0.188
50.6	Bedrock	0.20	0.20	0.20	0.20
Site #2 Northern New York					
0	Waste	0.348	0.347	0.149	0.097
6.4	Waste	0.282	0.224	0.11	0.079
14.0	Waste	0.203	0.222	0.104	0.096
21.64	Waste	0.189	0.238	0.130	0.126
29.26	Glacial Till	0.212	0.248	0.186	0.175
36.0	Bedrock	0.20	0.20	0.20	0.20
Site #3 Western Massachusetts					
0	Ash/Waste	0.309	0.418	0.129	0.12
4.6	Ash/Waste	0.282	0.118	0.094	0.089
11.0	Waste	0.228	0.197	0.083	0.081
15.2	Waste	0.251	0.347	0.149	0.145
19.8	Sand	0.335	0.292	0.169	0.18
24.4	Sand	0.352	0.243	0.178	0.202
32.0	Sand	0.323	0.202	0.173	0.189
39.6	Sand	0.301	0.178	0.130	0.172
47.2	Sand	0.263	0.176	0.178	0.164
54.8	Sand	0.225	0.194	0.185	0.168
62.5	Bedrock	0.20	0.20	0.20	0.20

The second characteristic of the seismic response evaluated is the response spectra for the cover system, liner system, and bedrock, as presented in Figure 4. The plots represent the spectral response of Site #1 to all four earthquake motions. The bedrock acceleration response spectrum defines peak responses of damped single-degree-of-freedom oscillators when subjected to a given earthquake acceleration time history. The primary observation from review of Figure 4 is that the west coast motions showed amplified spectral accelerations and damped spectral accelerations for the east coast motions for the fundamental period ranges of the landfills. Calculated fundamental periods of the landfills are in the range of 0.8 to 1.0 seconds. Another observation from reviewing the plots is that the west coast motion with the largest frequency content (Loma Prieta,1989) showed the largest amplification. As frequency content of the earthquake motions increased, the amplification decreased and became damped for the two east coast events.

A

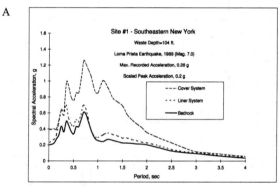

B

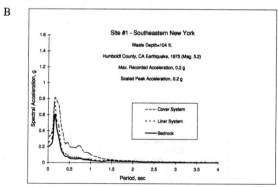

C

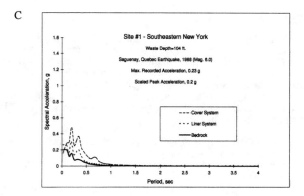

D

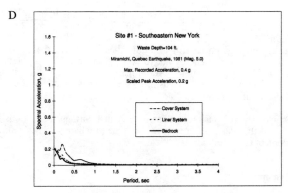

Figure 4. Spectral Accelerations for Site 1.

Discussion of Results Compared to Other Investigators

The results of this paper are in agreement with the findings of other investigators. Research by Kavazanjian et al. (1995) supports the author's findings that there is an amplification of west coast motions and a damping of east coast motions. Klimkiewicz (1995) reports that accelerations from east coast seismic events, which are typically high in frequency content, are damped throughout landfill sites. Additionally, Repetto et al. (1993) and Singh and Sun (1995) report that municipal solid waste has inherently moderate to strong energy absorption characteristics especially for high frequency motions.

Conclusions

The EPA regulations governing municipal solid waste require a seismic analysis of landfills that are located within seismic impact zones. The regulations include a requirement for the maximum earthquake accelerations to be used in the analysis, without identifying additional earthquake descriptors. The purpose of this paper was to investigate the impacts of one of the earthquake descriptors, frequency content, on the seismic response of three landfills in the northeastern U.S. The analysis was conducted with WESHAKE and four distinct earthquake motions.

The results of the analysis exhibit a clear distinction between the seismic response of the landfills for various input motions. As reported, the low frequency west coast motions tend to be amplified as they propagate through the waste mass whereas the high frequency east coast motions tend to be attenuated as they propagate through the waste mass. Although the input bedrock motions were all scaled to 0.2g, the landfill responses were quite different. This leads to the conclusion that frequency content has a much larger role in landfill response than does peak bedrock acceleration. The impact of this varied response is that ultimate factors of safety determined from a slope stability analysis could be very different. For a given landfill in the northeastern U.S., the use of a west coast motion could cause a practitioner to convey an instability problem, when in reality the landfill may be stable. Additionally, a displacement analysis for a landfill in the northeastern U.S. using a west coast motion could easily overpredict the strains anticipated.

An important conclusion of this study is that earthquake motions should be selected on the basis of regional earthquake characteristics, since the frequency content of the motion is critical.

Acknowledgments

The authors would like to gratefully acknowledge the following individuals for their assistance in preparing this paper. Dr. T. Stark, University of Illinois; J. Russo, Parsons Engineering Science, Inc.; J. Morrile, Syracuse University; and, V. DiCarlo, New York State Department of Environmental Conservation.

References

Algermissen, S.T., Perkins, D.M., Thenhaus, P.C., Hanson, S.L., and Bender, B.L. (1992). Probabilistic estimates of maximum acceleration and velocity in rock in the contiguous United States. *U.S. Geological Survey Open-File Report 82-1033*, 99 p.

Atkinson, G.M. (1987), "Implications of Eastern Ground Motion Characteristics for Seismic Hazard Assessment in Eastern North America," Proc. *Symposium on Seismic Hazards, Ground Motions, Soil-Liquefaction and Engineering Practice in Eastern North America*, Tuxedo, New York, NCEER Technical Report No. NCEER-87-0025.

Bray, J.D., Repetto, P.C., Augello, A.J., and Byrne, R.J. (1993). An overview of seismic design issues for solid waste landfills. *Proceedings*, Geosynthetics Research Inst. Conf., Drexel Univ., Philadelphia, PA, pp. 242-254.

Jacob, K., Armbruster, J., Barstow, N., and Horton, S. (1994). Probabilistic Ground Motion Estimates for New York: Comparison with Design Ground Motions in National and Local Codes. Proc. *Seismic Analysis and Design Considerations for Municipal Solid Waste Landfills*, Saratoga Springs, New York, pp. 1-10

Kavazanjian, E., and Matasovic, N. (1995). Seismic analysis of solid waste landfills. *Proceedings, Geoenvironmental 2000, New Orleans, LA, Feb. 1995, ASCE Geotechnical Special Publication No. 46, Vol. 2, pp. 1066-1080.*

Kavazanjian, E., Matasovic, N., Bonaparte, R., and Schmertmann, G.R. (1995). Evaluation of MSW properties for seismic analysis. *Proceedings*, Geoenvironmental 2000, New Orleans, LA, Feb. 1995, *ASCE Geotechnical Special Publication No. 46*, Vol. 2, pp. 1126-1141.

Klimkiewicz, G.C. (1995). Earthquake ground motions for landfill design. *Proceedings*, Geoenvironmental 2000, New Orleans, LA, Feb. 1995, *ASCE Geotechnical Special Publication No. 46*, Vol. 2, pp. 1097-1112.

Nuttli, O.W. (1981), "Similarities and Differences Between Western and Eastern United States Earthquakes, and their Consequences for Earthquake Engineering," Earthquakes and Earthquake Engineering: The Eastern United States, Vol1, Assessing the Hazard - Evaluating the Risk, J.E. Beavers, Ed., Ann Arbor Science Publishers, Inc., Ann Arbor, Michigan, pp. 25-51.

Repetto, P.C., Bray, J.D., Byrne, R.J., and Augello, A.J. (1993). Seismic Design of Landfills. Proc., Workshop on Research Priorities for Seismic Design of Solid Waste Landfills, University of Southern California, Los Angeles, CA, August 26, 1993, pp. 37-70.

Schnabel, P.B., Lysmer, J., and Seed, H.B. (1972). SHAKE: a computer program for earthquake response analysis of horizontally layered sites. Report EERC-72/12, *Earthquake Engineering Research Center*, Berkeley, CA.

Singh, S., and Murphy, B.J. (1990). Evaluation of the stability of sanitary landfills. In: Geotechnics of Waste Fills - Theory and Practice, *ASTM STP 1070*, pp. 240-258.

Singh, S., and Sun, J.I. (1995). Seismic evaluation of municipal solid waste landfills. *Proceedings*, Geoenvironmental 2000, New Orleans, LA, Feb. 1995, *ASCE Geotechnical Special Publication, No. 46*, Vol. 2, 1081-1096.

Sykora, D.W. (1987). Examination of existing shear wave velocity and shear modulus correlations in soils. Misc. Paper Gl-87-22, *U.S. Army Engineers Waterways Experiment Station*, Vicksburg, MS.

U.S. Environmental Protecton Agency. (1995). RCRA Subtitle D (258), Seismic Design Guidance for Municipal Solid Waste Landfill Facilities, pp. 1-143.

U.S. Environmental Protection Agency. (1991). Title 40, Part 258, Criteria for Municipal Soils Waste Landfills, Code of Federal Regulation, pp. 355-361.

University of the State of New York. (1991), Geology of New York, A Simplified Account, Educational Leaflet No. 28, pp. 1-284.

In-Situ Testing Methods For Dynamic Properties Of MSW Landfills

W. N. Houston[1], S. L. Houston[2], J. W. Liu[3], A. Elsayed[4] and C. O. Sanders[5]

Abstract

Cost-effective techniques have been developed for measuring P and S-wave velocities and in-situ shear strength for landfill waste materials. These techniques include both surface profiling and downhole measurements to capture the typical velocity inversion which accompanies a compacted soil cover. The measurement scheme is basically a simplified tomography survey which characterizes the material within a large cone around the borehole. A computer code is being developed to fit the data to a statistical model for each cone. Using a least-squares fit the code will automatically iterate to mean values and quantify the local variability (on a scale of 5 to 20 m). Variability across the landfill is assessed by repeating the measurements at several locations.

Static shear strength of landfill materials is also being measured with a large-scale (1.22m x 1.22m) in-situ direct shear box. Because it is common practice to use a percentage of the static strength as the seismic strength in obtaining yield accelerations, these static strengths are useful for earthquake analyses. Attempts will be made to correlate these in-situ shear strengths with wave velocities and moduli at the same locations when more in-situ strength data are available.

Introduction

Analysis of landfill stability under seismic loading is required by US Federal regulations in seismic impact zones. However, very little data is available regarding the mechanical properties of municipal solid waste (MSW), especially dynamic properties. The subject of this paper is the determination of mechanical properties of MSW by field testing. The two test methods employed are: (1) seismic wave

[1,2]Professor, Department of Civil Engineering, Arizona State University, Tempe, AZ 85287-5306
[3,4]Research Assistant, Department of Civil Engineering, Arizona State University.
[5]Assistant Professor, Department of Geology, Arizona State University, Tempe, AZ 85287.

measurements, using both surface profiling and downhole measurements, and (2) large-scale in-situ direct shear tests. This paper represents a progress report on ongoing research and development of tools and methodologies for field testing of difficult-to-sample materials. The emphasis in the research program has been on development of low-cost apparatus and techniques that can be used to obtain good data.

Description Of Approach

Seismic wave measurements were chosen as one of the techniques because a large volume of material can be characterized at relatively low cost. As the first phase of the seismic velocity component, the technique of tomography was studied in great detail. In brief, tomography consists of characterizing a mass of material by sending hundreds or perhaps even thousands of seismic rays through it. This is usually done by placing energy sources and detectors (geophones) at many points on the outer boundaries of the mass. The mass is subdivided into modules, and a computer code is used to solve a set of simultaneous equations, which then gives values for the velocity for each module. As normally employed, tomography would use as many modules as possible, while ensuring that the number of unknowns is less than or equal to the number of equations. In this format, there are few, if any, redundant measurements. At the completion of the tomographic study, mean values and variability could be assessed by looking at the individual values for the modules.

The approach chosen in this study for landfill application is somewhat different, and can be referred to as simplified tomography. Use of a very small number of modules has been selected and this choice results in considerable redundancy in the measurements. This redundancy is used to compute mean values and standard deviation as a measure of local variability. Quantified variability as well as best estimates are needed for risk analysis.

Figure 1 is a schematic of the testing configuration. The geophones on the ground surface are nominally for surface profiling, in the configuration represented by Figure 1, the surface geophones typically receive only direct rays as first arrivals and the redundancy of measurements allows for computation of means and standard deviations. The downhole geophones receive rays from all energy source locations. When conducting seismic surveys it is customary to move both geophones and energy sources from time to time. The procedure adopted for these studies is to keep the geophones in place and move only the energy source locations. An exception to this rule is the downhole geophones, which are placed at a maximum of two locations in the borehole.

Figure 1(b) also shows a possible ray path from an outer energy source location to a down-hole geophone. The velocity within the soil cover is assumed constant so the ray path is a straight line in the soil cover. Due to the lower velocity in the upper refuse, the ray refracts downward rather than upward. Because of the typical gradual increase in velocity with depth for refuse, the possible ray path in the refuse curves as shown. The position and shape of the first arrival ray path can be

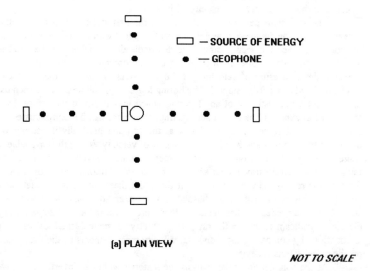

(a) PLAN VIEW

NOT TO SCALE

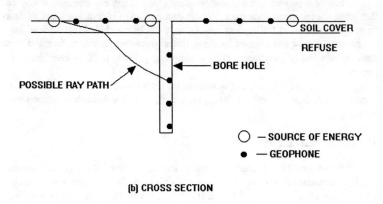

(b) CROSS SECTION

Figure 1. Schematic of Seismic Velocity Testing Configuration

estimated from the ground geometry, ratio of velocity in the soil cover and upper refuse, and the rate of increase velocity with depth.

The analysis is performed in stages. For the first stage, the surface profiling data is taken as the database. These data establish a first estimate of the velocities of the soil cover. For the second stage, the downhole data is added to the database. The downhole data includes the boring log and the downhole travel time; this data also includes a measure of velocities of the soil cover. Therefore, these velocities can be updated as a part of stage 2. The boring log and downhole data are then used to construct a first estimate of any layering which may occur in the refuse. In the absence of any indications of distinct layering, the refuse can be approximated as a single layer with some value of velocity at the top and gradually increasing with depth. At the end of stage 2, a preliminary plot of velocity vs. depth is available. At stage 3, the data from the outer energy sources are added to construct best- estimate ray paths and then the travel times are used to provide additional values of velocities for both the refuse and the soil cover. At this point the database is complete and a best-fit velocity profile can be established. However, an additional iteration back to stage 2 may be needed, in that the last velocity profile (at the end of stage 3) might dictate a significant change in the ray path geometry. In stage 5, the local variability is quantified by comparing the individual velocity measurements to the final best-fit profile to get standard deviations.

At the completion of the analysis the material within an inverted cone, with the borehole at the center, has been characterize. Variability across the landfill is assessed from several boreholes across the landfill.

Future improvements to the methodology are likely. For example, it may be that when the computer code for automatic data reduction is completed and all data from the first phase of testing is in final form, it will be concluded that minimal loss in accuracy will be experienced by using fewer geophones at the surface and that three surface line of geophone at 120° is almost as good as the four lines shown in Figure 1.

As a part of research and development, testing was done by the cross-hole, uphole, and down-hole methods. Based on comparisons, the down-hole method was selected.

The Downhole Test Method

In the downhole method, the energy source is located at the ground surface and the geophones at various elevations in a drill hole. Strong shear waves can be generated by excavating a small trench at the surface near the borehole and installing a metal plate at each end of the trench. Strong compression waves can be generated with vertical blows from a large hammer or a water-filled tank which is filled in the field and hoisted into the air and dropped. Advantages of downhole over cross-hole or uphole testing include: (a) Compared to cross-hole, only one borehole is needed. (b) Shear wave arrivals are easily detected by using a reversible bi-directional energy source. (It is difficult to devise an energy source which is both powerful and reversible that can be discharged from within the borehole). (c) At the surface it is more convenient to employ a powerful energy source. (d) Vertically propagating shear waves are generated and measured, and these correspond to earthquake loading with shear waves.

An open drill hole was chosen because interpretation of downhole velocity data can be very difficult if the bore hole casing causes the "tube waves" problem. The "tube waves" means that there is contamination of wave arrivals through the medium with reflected wave arrivals in the casing when medium velocities are lower than that of the casing.

A three dimensional seismic receiver assembly has been developed (Figure 2). The assembly consists of three standard geophones mounted at mutually perpendicular angles. This configuration allows polarized waves to be detected. The assembly's brace can be altered by the users. Current braces can be used in holes ranging from 10 cm (4 inches) to 36 cm (14 inches) in diameter. The rope connected to the top of the apparatus is for pulling the sensors out of the hole and guiding the pipe back onto them to move them up or down in the hole. The geophones are placed downhole using 1-1/2" schedule 40 PVC pipe. The pipe comes in 3.05m(10ft) lengths connected together using 0.305m(1ft) long sleeves of 2" tubing. A vertical mark is placed along the length of the PVC pipes that lines up with a known alignment of the geophones, so the orientation of the geophones is known.

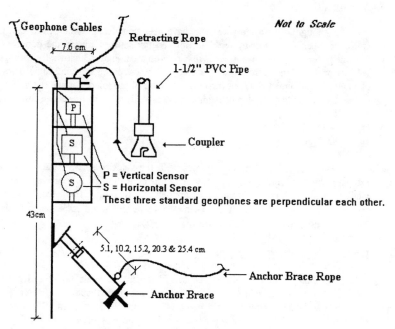

Figure 2. 3-D Seismic Receiver Assembly

Site Conditions

The Northwest Regional Landfill Facility (NWRLF) is located in rural north western Maricopa County, Arizona. It is a municipal solid waste (MSW) landfill. The landfill accepts MSW, inert construction debris, and landscaping debris. Liquids and hazardous waste materials are not accepted for disposal. SCS ENGINEERS performed a geotechnical exploration program at the NMRLF site in December 1993: seven test borings were drilled to depths ranging from 15.24m(50 ft) to 50.29m(165 ft) below the existing ground surface. Groundwater was not encountered in any of the borings (SCS ENGINEERS,1994). Approximately 711 acres of the site are planned for land disposal of solid wastes. The site has been in continuous operation since December 1988. The minimum expected life of the facility is 50 years. The active landfill area is in Phase I which comprises approximately 106 acres of land located at the northwestern portion of the site. Five borings, 76.2m (250ft) apart each other, were drilled for the purpose of measuring dynamic material properties of the refuse fill. Depths of the borings in this study were from 8.5 m (28 ft) to 9.1 m (30 ft). Greater depths were not drilled due to landfill management concern.

Test Results

Although all field measurements have been completed, some of the data is still under analysis, especially the data from energy sources around the periphery. For this paper only the down hole results are presented. Figure 3 shows the average velocity profiles derived from testing at five bore holes spaced at 76m(250 ft). These plots clearly show the velocity inversion at the base of the soil cover and the gradual increase in velocity with depth. Additional data derived from these velocities are presented in Table 1.

Error Analysis

The accumulated error associated with the velocity survey results can be estimated by considering the accuracy of the identification of first arrival time as well as the accuracy of the distance measurements. In this study, the percent error in the travel time and distance were estimated to be ±5% and ±4% respectively. Therefore, the maximum velocity error is about ±9%.

Large Scale In-Situ Direct Shear Tests

An analysis of the seismic stability of landfill materials requires dynamic strength values for the refuse. It is common practice in earthquake engineering to estimate the dynamic strength as a percentage of the static strength. Thus static strength from large-scale direct shear tests would be more or less directly usable in a seismic stability analysis.

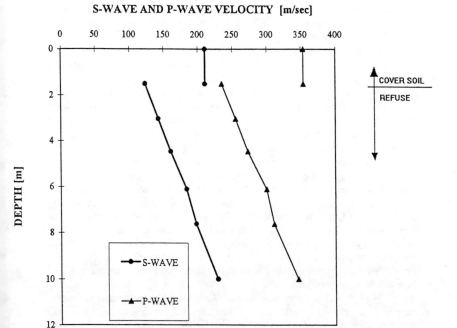

Figure 3. Predicted Velocity For MSW Landfill

Table 1. Dynamic Moduli And Poisson's Ratios
For MSW Landfill From This Study

Depth (m)	P Wave Velocity (m/sec)	S Wave Velocity (m/sec)	$\frac{V_p}{V_s}$	Poisson's Ratio	Shear Modulus (MPa)	Young's Modulus (MPa)
1.52	235	124	1.89	0.30	16.1	42.0
3.05	255	143	1.78	0.27	21.4	54.2
4.47	273	161	1.70	0.23	27.0	66.5
6.10	300	184	1.63	0.20	35.0	84.1
7.62	311	198	1.57	0.16	40.8	94.8
10.0	346	229	1.51	0.11	54.7	121.3

To date, two in-situ direct shear tests have been performed at the NWRLF at seismic velocity measurement locations. The direct shear apparatus and testing procedure will be described in detail, with sketches and drawings, in future publications under preparation. Due to space limitations, only a short verbal description will be given here.

The box has three compartments which are stacked, the internal dimensions are 1.22m × 1.22m, and the height of each compartment is about 0.76m. The framework for each compartment is of welded steel with angle iron for bracing. Each compartment is split in half for easy lifting and transporting. The two halves are bolted together at the site. There are no tops or bottoms to these compartments. The framework has slots on four sides, into which beams can be dropped to form walls. The beams for the bottom compartment are of hollow Al tubing with rectangular cross-section, because of high lateral loads in the bottom compartment. Wooden 2×6's are used for beams in the upper compartments. A 25,000 lb hydraulic jack is used to apply the shear load, and is configured to apply a tensile force. A heavy chain around the base of the lower compartment is attached to one end of the jack and the other end of the jack is attached to a threaded steel rod about 4m long, which is secured to a small portable wooden wall that provides passive resistance. The wooden wall is placed in a trench which is perpendicular to the shear force direction. A backhoe excavates the trench for the wooden wall and a perpendicular trench for the threaded rod, and then covers the threaded rod and mounds up soil in front of the wall for more passive resistance. The threaded rod is housed in a PVC pipe which can be sacrificial.

The bottom compartment is assembled and placed on the ground surface (or on a bench at greater depth formed by the backhoe). Then the backhoe excavates a trench on all four sides of the box and the field technicians assist with hand trimming so that the box slides down over the cube of undisturbed material. When the cube occupies about half the volume of the first compartment the downward movement of the box is halted and the backhoe finishes filling the box with loose soil. Buckets of known volume are used to collect samples of the loose soil, so the density of the fill can be estimated and the normal stress can be calculated. After the box is filled and leveled, the jack is attached to the chain and the reaction wall and first stage of loading is begun. This stage is illustrated in Figure 4 which shows the results of the two tests at the NWRLF, tests C1 and C2. For the first stage the load is increased to the nearest mm by attaching a meterstick to each side of the box and noting movement relative to a reference stake driven in behind the box. When the shear stress levels off, the second stage is begun by placing the second compartment on top of the first and filling it with soil to obtain additional normal stress. After a peak (or leveling off) is reached for stage 2, stage 3 is commenced.

The test interpretation used to date is conventional and the results are shown in Figure 5. By the time stage 2 and 3 are reached, the movement has been significant and it is reasonable to assume that at least some of the cohesion has been destroyed. A commonly used assumption is that φ will be relatively unaffected by this movement. Therefore, the dashed lines in Figure 4 represent a good approximation of c and φ for first stage loading, and it is these values that are

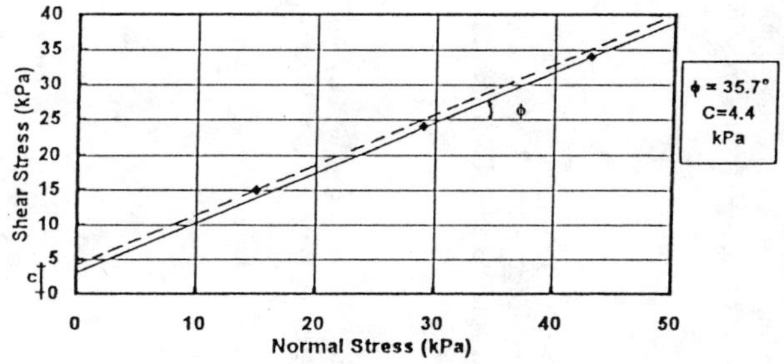

Figure 5(a) Shear Stress vs. Horizontal Displacement; (test # C1)

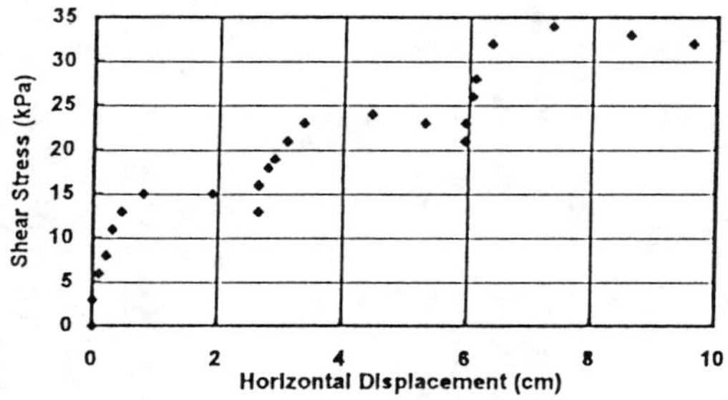

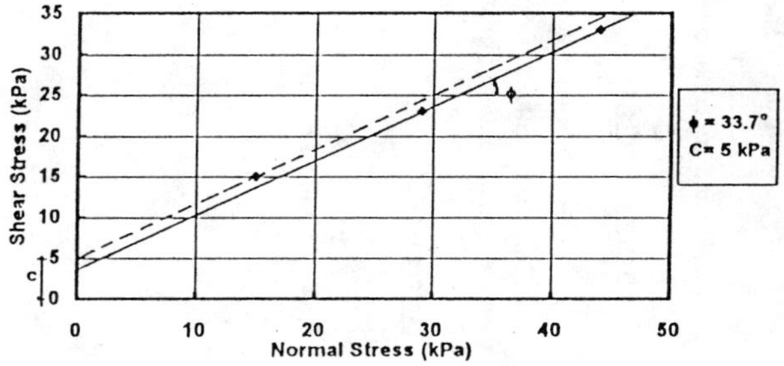

Figure 5(b) Shear Stress vs. Horizontal Displacement; (test # C2)

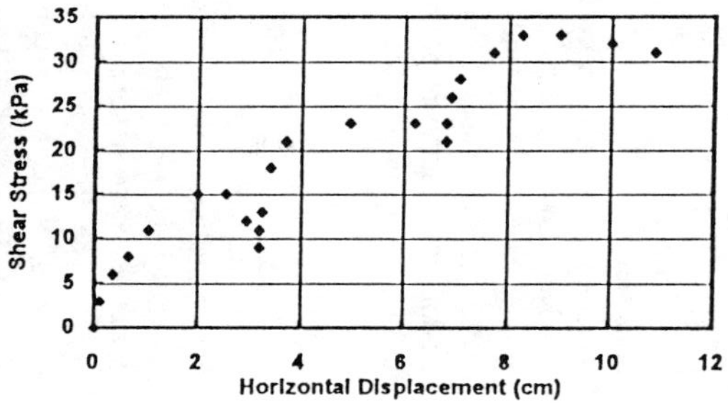

Figure 4(a) In-Situ Large-Scale Direct Shear Test; (test # C1)

$\phi = 35.7°$
$C = 4.4$ kPa

Figure 4(b) In-Situ Large-Scale Direct Shear Test; (test # C2)

$\phi = 33.7°$
$C = 5$ kPa

9

reported. It is remarkable that tests at two separate locations in the landfill gave such closely agreeing results. The stress state in and around the shear box is currently being subjected to finite element analyses and other than conventional test interpretation may sometime be recommended.

Large-scale in-situ direct shear tests are not a new concept. What is considered to be new and innovative about these tests is that the apparatus and methodology have been modified through trial and error to arrive at an efficient, relatively low cost test. Several of these tests have been performed in cobbly gravels, and two in the landfill. Two technicians together with the backhoe operator have always been able to perform one test per day, and sometimes two tests. At the landfill, two tests were performed in one day even though it was the first time for testing refuse material.

Conclusions

Results from studies to date indicate that characterization of landfill mechanical properties at reasonable cost is an achievable goal. The seismic velocity measurements and the large-scale in-situ direct shear test appear to be suitable tools for these determinations.

The downhole test method was compared to cross hole and up hole test methods, although this comparison was not shown in this paper due to space limitations. It was concluded the ratio of data obtained to cost of testing was minimum for downhole testing. Therefore the downhole test was adopted.

The in-situ direct shear testing proceeding smoothly and it was possible to perform two tests on the refuse on the first day of testing. The test results appear reasonable and are conducted at low cost.

Acknowledgments

This research is based upon work suported by NSF under Grant No. CMS-9301285. The authors would like to acknowledge Keith A. Johnson, Enamul Hoque, and a number of geotechnical graduate students at ASU for their assistance.

References

SCS ENGINEERS, Aquifer Protection Permit Application Document, Northwest Regional Landfill, Maricopa county, Arizona, Book 1, April 19,1994.

Seismic Response of The Operating Industries Landfill

I. M. Idriss [1], Fellow, ASCE, Gregg Fiegel [2], Martin B. Hudson [3], Associate Member, ASCE, Peter K. Mundy [4], Member, ASCE, and Roy Herzig [5]

Abstract: The Operating Industries, Inc. (OII) Landfill is located about 10 miles east of Los Angeles within the City of Monterey Park, California. Shortly after the 1987 Whittier-Narrows earthquake, two strong motion recording stations were placed at this landfill. One station was placed at the toe of the landfill in an area that had been filled with granular fill over natural soils. The other station was placed at the top of the landfill. Since their installation, the two instruments have recorded ground motions during 40 earthquakes. The purpose of this study was to utilize motions recorded during four of these earthquakes to derive modulus reduction and damping curves for the landfill waste using both two-dimensional and one-dimensional seismic response procedures. These curves were derived based on back-calculating the motions at the top of the landfill for these four earthquakes and obtaining reasonably good matches between the calculated and the recorded motions for all four earthquakes. The results of one dimensional analyses did not provide as good a match as those obtained using the two-dimensional analyses. The modulus reduction curve derived in this study for waste exhibits far less modulus reduction, in the large strain range, than curves suggested by others.

Introduction

Landfills have been, and continue to be, built in highly seismic regions. While many of these landfills have been subjected to various levels of shaking during past earthquakes, only one landfill is equipped with strong motion instruments capable of recording the level of shaking experienced by the landfill. Two strong motion

[1] Depart. of Civil & Environmental Engineering, University of California, Davis 95616
[2] CDM Federal Programs Corporation, Walnut Creek, California
[3] Law / Crandall, Inc., Los Angeles; formerly, with CDM Federal Programs Corporation
[4] CDM Federal Programs Corporation, Walnut Creek, California
[5] US Environmental Protection Agency, San Francisco

instruments were placed at the Operating Industries, Inc. (OII) Landfill shortly after the October 1, 1987 Whittier-Narrows earthquake. These strong motion instruments recorded the motions generated during 40 earthquakes over the past seven years.

The purpose of this paper is to utilize four sets of the recordings obtained at this site to assess the performance of this landfill during earthquakes and to establish material characteristics, such as modulus reduction and damping curves, for waste material at landfills. Additional considerations, including the use of one- or two-dimensional analysis procedures are also discussed in this paper.

Site Conditions

The OII Landfill is located about 10 miles east of Los Angeles within the City of Monterey Park, California. The landfill covers approximately 190 acres and is divided by the Pomona Freeway (Highway 60). The landfill area south of the freeway, known as the South Parcel, covers approximately 145 acres and is the major part of the landfill. The layout of this parcel and the location of the strong motion instruments (designated as Seismic Stations SS-1 and SS-2) are shown in Fig. 1. The ground surface adjacent to Station SS-1 is at about elevation 155.9 m (511.4 ft) and that adjacent to Station SS-2 is at about elevation 191.9 m (629.6 ft); these elevations are based on a survey completed in late 1994.

Filling operations began at the OII Landfill in 1948. Residential and commercial waste, industrial wastes, liquid wastes, and various hazardous wastes were disposed at the landfill throughout its operating life. The landfill operators stopped accepting hazardous liquid wastes in January 1983, and all liquid wastes in April 1983. Land-filling operations ceased in October 1984. The landfill is in the process of final closure.

Several geotechnical investigations have been conducted at this site. The latest of these investigations consisted of drilling and sampling of one boring adjacent to each seismic station. The shear and compression wave velocities at each location were then measured using various geophysical techniques.

A section across the landfill, and including the two seismic stations, is presented in Fig. 2. This section is based on the results of the previous geotechnical investigations completed over the past 15 years. The subsurface conditions at the location of Seismic Station SS-1 (designated as the Toe Station) consist of approximately 30.5m (100 ft) of fill underlain by natural material consisting mostly of the Pico Unit, which is composed of alternating sandstones, sandy shales, clayey shales and siltstones with varying degrees of weathering. The subsurface conditions at the location of Seismic Station SS-2 (designated as the Top station) consist of approximately 7.3 m (24 ft) of soil cover underlain by waste to a depth of about 99

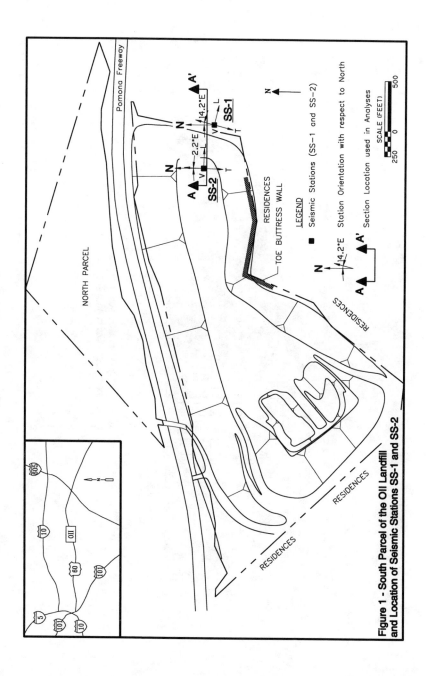

Figure 1 - South Parcel of the Oil Landfill and Location of Seismic Stations SS-1 and SS-2

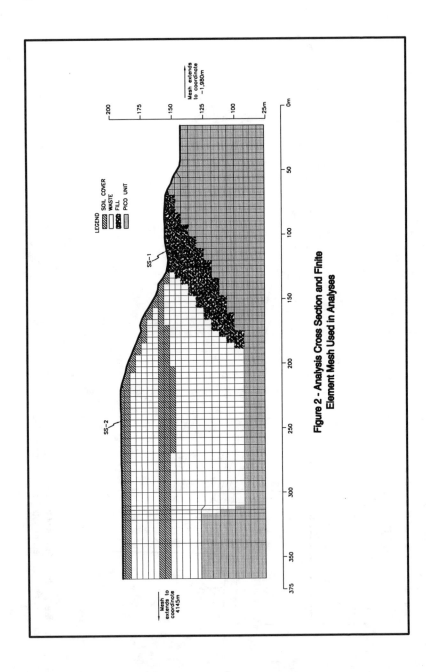

Figure 2 - Analysis Cross Section and Finite Element Mesh Used in Analyses

m (325 ft) with a 13.7 m (45 ft) thick layer of soil cover at a depth of about 30.5 m (100 ft) as shown in Fig. 2.

The shear wave velocities measured adjacent to the Toe Station are presented in Fig. 3. The shear wave velocities at this location were measured using the OYO suspension logger and the spectral-analysis-of-surface-waves (SASW) testing procedure. The best estimate velocities at the Toe Station, established based on these measurements, are also shown in Fig. 3.

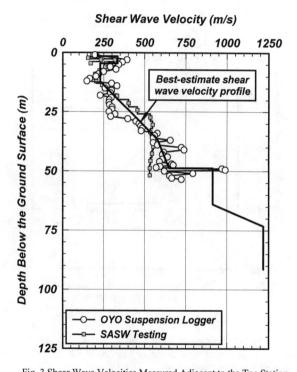

Fig. 3 Shear Wave Velocities Measured Adjacent to the Toe Station

The shear wave velocities measured adjacent to the Top Station are presented in Fig. 4. The shear wave velocities at this location were measured using the OYO suspension logger, the spectral-analysis-of-surface-waves (SASW) testing procedure, and the downhole logging procedure. The best estimate velocities at the Top Station, established based on these measurements, are also shown in Fig. 4.

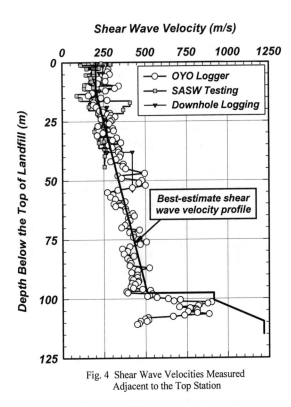

Fig. 4 Shear Wave Velocities Measured
Adjacent to the Top Station

The shear wave velocities below a depth of about 100 m (325 ft) were not used in establishing the best estimate velocity values adjacent to the Top Station because of sloughing around the casing at the bottom of the boring.

A shear wave velocity of 1220 m/s (4000 ft/sec) was assigned in the underlying rock half-space at both locations as shown in Figs. 3 and 4.

The average value of the maximum shear wave velocities (i.e., velocities measured at very small strain levels using the geophysical techniques indicated above) of the soil profile adjacent to the Toe Station is approximately equal to 455 m/s (1500 ft/s); the corresponding value of the waste/soil profile adjacent to the Top Station is approximately equal to 365 m/s (1200 ft/s). The average value was calculated by adding the travel time over depth increments of about 10 m and then dividing into the total thickness of the profile.

Using the average value, v_s, of the maximum shear wave velocities of 455 m/s (1500 ft/s) and a total thickness, H, of 73 m (240 ft) indicates a minimum fundamental period, T, of about 0.64 s ($T = 4H/v_s$) for the profile adjacent to the Toe Station. The corresponding numbers for the profile adjacent to the Top Station are: $v_s = 365$ m/s (1200 ft/s), $H = 109$ m (360 ft) and $T = 1.2$ s. These values of the shear wave velocities are used below as part of examining the characteristics of the earthquake ground motions recorded at the Toe and at the Top Stations.

Lower and upper range shear wave velocities were also used by allowing a change, from the best estimate values, ranging from about 10 percent near the bottom of the profile to about 30 percent near the top of the profile.

The fill and the natural materials in the upper 6.1 m (20 ft) were assigned a total unit weight of 18.86 kN/m^3 (120 pcf); below this depth this unit weight was increased to 19.65 kN/m^3(125 pcf). The soil cover was assigned a total unit weight of 17.29 kN/m^3 (110 pcf). The total unit weight assigned to the waste varied linearly from 11 kN/m^3 (70 pcf) at the ground surface to 14.15 kN/m^3 (90 pcf) at a depth of 39.6 m (130 ft), below which it was kept at this value.

Earthquake Ground Motions Recorded at OII

Ground motions generated during 40 earthquakes were recorded by the strong motion instruments placed at these two seismic stations. The ground motions recorded during eighteen of these earthquakes were fully processed and are available in digitized form (USEPA 1995). The magnitude of these earthquakes ranged from 4.1 to 7.3, and the distance to the source of energy release ranged from about 10 km to 160 km.

Ground Motions Used in this Study: The ground motions recorded during four of these earthquakes were used in this study. The name, date, magnitude, style of faulting, and closest distance to the OII site for each earthquake are listed below:

Earthquake	Date	M	Style of Faulting	Dist. (km)
Pasadena	03 December 1988	5.0	Reverse	14
Landers	28 June 1992	7.3	Strike slip	140
Northridge	17 January 1994	6.7	Blind Thrust	44
Northridge Aftershock	19 January 1994	4.5	Blind Thrust	45

Values of the horizontal (east and west components) and the vertical peak acceleration, peak velocity and peak displacement together with the duration for the motions recorded at the Toe Station during each earthquake are listed below:

| Earthquake | Component | Values of Ground Motion Parameters Recorded at the Toe Station | | | |
		pga (g)	pgv (cm/s)	pgd (cm)	Duration (sec)
Pasadena	North	0.150	6.45	0.39	3.4
	East	0.218	9.70	0.62	3.8
	Vertical	0.100	3.64	0.18	4.7
Landers	North	0.029	11.57	18.25	46.7
	East	0.042	10.87	6.85	41.4
	Vertical	0.021	4.05	3.00	49.4
Northridge	North	0.224	12.06	2.02	10.5
	East	0.258	19.49	2.38	10.9
	Vertical	0.151	6.93	1.59	14.0
Northridge	North	0.027	0.97	0.05	5.7
(aftershock)	East	0.043	1.59	0.08	5.2
	Vertical	0.034	1.27	0.06	5.0

The corresponding values of the parameters recorded at the Top Station during each earthquake are listed below:

| Earthquake | Component | Values of Ground Motion Parameters Recorded at the Top Station | | | |
		pga (g)	pgv (cm/s)	pgd (cm)	Duration (sec)
Pasadena	North	0.118	4.86	0.40	4.6
	East	0.091	6.33	0.51	3.3
	Vertical	0.070	4.07	0.26	4.3
Landers	North	0.070	17.95	19.17	43.1
	East	0.100	23.19	7.43	39.3
	Vertical	0.024	5.68	3.16	44.5
Northridge	North	0.205	16.76	2.73	19.3
	East	0.254	27.26	5.37	16.3
	Vertical	0.139	9.20	1.79	16.9
Northridge	North	0.020	0.82	0.06	6.1
(aftershock)	East	0.031	1.92	0.15	5.0
	Vertical	0.018	1.02	0.06	4.4

In th above listings, M is moment magnitude, (pga) is peak ground acceleration, (pgv) is peak ground velocity, (pgd) is peak ground displacement, and duration for each accelerogram was obtained as the time required for the 5% to the 95% build-up of the Husid plot for that accelerogram. This definition of duration was initially proposed by Trifunac and Brady (1975) and has been used by other researchers over the past 20 years.

The peak horizontal accelerations of the motions recorded at these two stations, during the four earthquakes under consideration, are presented in the upper part of Fig. 5. This part of Fig. 5 shows the peak horizontal accelerations recorded at the top of the landfill plotted versus the peak horizontal accelerations recorded at the Toe Station. The values shown in this part of the figure indicate that the landfill amplified the peak acceleration during the Landers earthquake (magnitude 7.3 at a distance of about 140 km), de-amplified the acceleration during the Pasadena earthquake (magnitude 5 at about 14 km), and resulted in practically the same peak horizontal accelerations during the Northridge earthquake (magnitude 6.7 at about 44 km) and the Northridge aftershock (magnitude 4.5 at about 45 km).

The corresponding peak horizontal velocities are shown in the middle part of Fig. 5. The peak velocities of the motions recorded at the Top Station are significantly larger than those of the motions recorded at the Toe Station during the Landers and during the Northridge earthquakes. The velocities of the motions recorded during the Pasadena earthquake at the Top Station are smaller than those of the motions recorded at the Toe Station. The values obtained during the Northridge aftershock are almost the same at both stations.

The values of peak horizontal displacements are presented in the lower part of Fig. 5. The peak horizontal displacements of the motions recorded at the two stations are almost the same except for the east component of the recording at the Top Station obtained during the Northridge earthquake.

The Fourier spectrum for the accelerogram of the east component of motion recorded at each station during the Pasadena, the Landers and the Northridge earthquakes were calculated to examine the frequency characteristics of these motions. The Fourier spectra were smoothed by using a three point central difference smoothing procedure, which was repeated eleven times.

The resulting un-smoothed and smoothed Fourier spectra for the motions recorded at the Top and at the Toe Stations during the Pasadena earthquake are presented in Fig. 6. The largest peak in the smooth spectrum for the motion recorded at the Top Station occurs at a frequency, $f \approx 2$ Hz, which is probably related to the source for this magnitude 5 earthquake. Note that the smoothed spectrum for the motion recorded at the Toe Station also shows a peak at this frequency. The apparent peak at $f \approx 0.88$ or at $f \approx 1.1$ Hz is more likely to be related to the site characteristics of

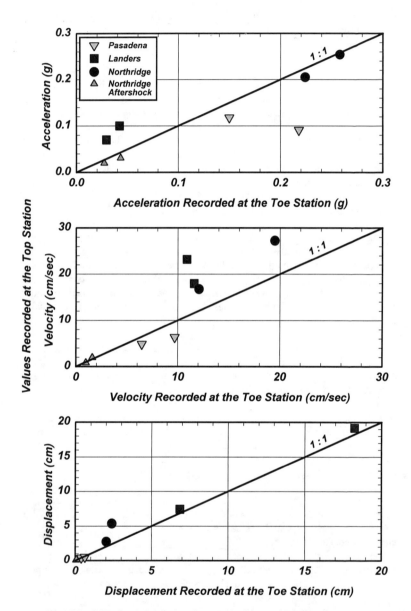

Fig. 5 Peak Horizontal Accelerations, Velocities and Displacements
of Motions Recorded at the Top and at the Toe Stations

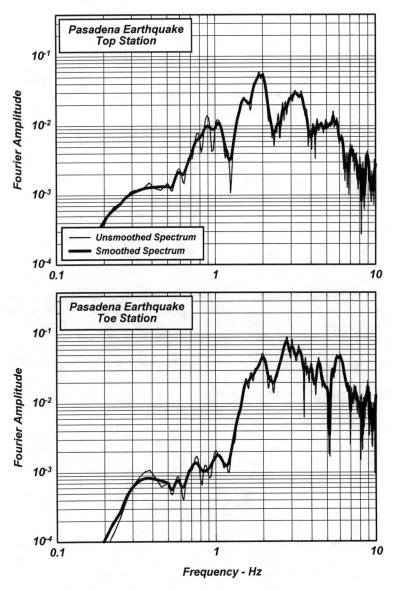

Fig. 6 Fourier Amplitudes for Motions Recorded at the Top
and at the Toe Stations During the 1988 Pasadena Earthquake

the profile adjacent to the Top Station. Similarly, the peak at $f \approx 0.75$ Hz is probably related to the site characteristics of the profile adjacent to the Toe Station. The corresponding "average" shear wave velocity for each profile would be given by $v_s = 4Hf$, $(f = 1/T)$ or $v_s \approx 380$ m/s (1250 ft/s) for the profile adjacent to the Top Station and $v_s \approx 220$ m/s (720 ft/s) for the profile adjacent to the Top Station.

These values of the average shear wave velocity may be considered to represent a first order approximation of the "strain-compatible" shear wave velocity, $(v_s)_{avg}$, for the profile under consideration.

An order of magnitude estimate of the shear strains that may have been induced during an earthquake can be obtained by dividing the peak horizontal velocity, pgv, by $(v_s)_{avg}$. Thus, the following values of maximum shear strain, γ_{max}, and equivalent uniform shear strain, γ_{unf}, would be obtained for the profile adjacent to each station during the Pasadena earthquake:

Location	pgv cm/s	$(v_s)_{avg}$ m/s	γ_{max} percent	γ_{unf} percent
Top Station	6.3	380	0.017	0.007
Toe Station	9.7	220	0.044	0.018

Note that the pgv used above is the peak velocity of the east component of the recorded motions, and that the equivalent uniform shear strain was obtained by multiplying the maximum strain by a factor of 0.4, which is the value suggested in the SHAKE91 manual for use with a magnitude 5 earthquake.

The ratio of $(v_s)_{avg}$ divided by the average value of the maximum shear wave velocities, $(v_s)_{max}$, is equal to about 1.04 (i.e., essentially 1) in the waste / soil profile adjacent to the Top Station, which indicates that there was no reduction of the shear wave velocity in the profile at this location during this earthquake.

The ratio of $(v_s)_{avg}$ divided by $(v_s)_{max}$ is equal to about 0.48 in the soil profile adjacent to the Toe Station, which corresponds to $G/G_{max} = 0.23$. This modulus reduction value at a uniform strain of 0.018 percent is comparable to the lower range for sands published by Seed and Idriss (1970).

The Fourier spectra for the motions recorded during the Landers earthquake are shown in Fig. 7. The smoothed spectrum for the motion recorded at the Top Station indicates two main peaks (of almost identical amplitude); one peak is at a frequency of about 0.63 Hz and the other peak is at a frequency of about 0.88 Hz; the average would be about 0.75 Hz. The resulting $(v_s)_{avg}$ would then be about 330 m/s (1080 ft/s). The smoothed spectrum for the motion recorded at the Toe Station indicates a

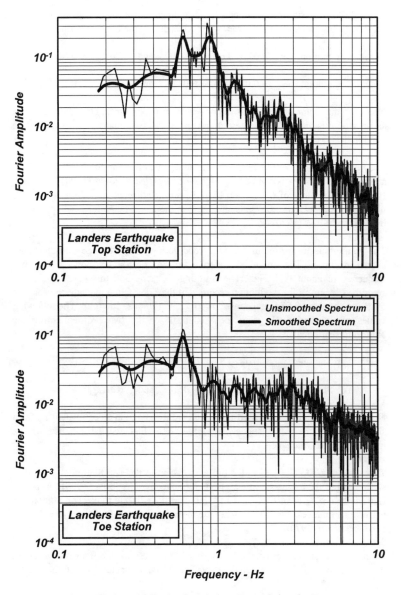

Fig. 7 Fourier Amplitudes for Motions Recorded at the Top
and at the Toe Stations During the 1992 Landers Earthquake

peak at a frequency of about 0.63 Hz; the resulting $(v_s)_{avg}$ would then be about 185 m/s (605 ft/s). Using the same procedure outlined above, the following values of γ_{unf} and G/G_{max} are obtained for the two profiles during the Landers earthquake:

Top Station: $\gamma_{unf} \approx 0.07 \times 0.63 \approx 0.044\%$ and $G/G_{max} \approx 0.81$

Toe Station: $\gamma_{unf} \approx 0.059 \times 0.63 \approx 0.037\%$ and $G/G_{max} \approx 0.17$

Note that the factor of 0.63 used to convert from maximum to equivalent uniform shear strain is the value suggested in the SHAKE91 manual for use with a magnitude 7.3 earthquake.

The value of $G/G_{max} \approx 0.81$ in the profile adjacent to the Top Station at a uniform strain level of 0.044% is comparable to that of clay used by Idriss (1990) for Young Bay Mud. The value of $G/G_{max} \approx 0.17$ in the profile adjacent to the Toe Station at a uniform strain level of 0.037% is smaller than the lower range for sands published by Seed and Idriss (1970).

The Fourier spectra for the motions recorded during the Northridge earthquake are shown in Fig. 8. The smoothed spectrum for the Top Station shows a peak at a frequency of about 0.78 Hz. The resulting $(v_s)_{avg}$ is 340 m/s (1120 ft/s). The smoothed spectrum for the motion recorded at the Toe Station indicates a peak at a frequency of about 0.58 Hz; the resulting $(v_s)_{avg}$ is 170 m/s (560 ft/s). The following values of γ_{unf} and G/G_{max} are then obtained for the two profiles during the Northridge earthquake:

Top Station: $\gamma_{unf} \approx 0.08 \times 0.57 \approx 0.046\%$ and $G/G_{max} \approx 0.85$

Toe Station: $\gamma_{unf} \approx 0.093 \times 0.57 \approx 0.05\%$ and $G/G_{max} \approx 0.14$

Note that the factor of 0.57 used to convert from maximum to equivalent uniform shear strain is the value suggested in the SHAKE91 manual for use with a magnitude 6.7 earthquake.

The value of $G/G_{max} \approx 0.85$ in the profile adjacent to the Top Station at a uniform strain level of 0.046% is slightly above that of clay used by Idriss (1990) for Young Bay Mud. The value of $G/G_{max} \approx 0.14$ in the profile adjacent to the Toe Station at a uniform strain level of 0.05% is smaller than the lower range for sands published by Seed and Idriss (1970).

Examination of the upper part of Fig. 8 suggests that a peak at a frequency of about 0.59 Hz may well be considered in evaluating the characteristics of the profile

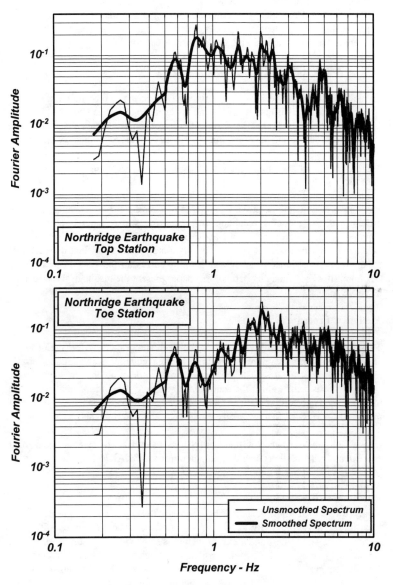

Fig. 8 Fourier Amplitudes for Motions Recorded at the Top
and at the Toe Stations During the 1994 Northridge Earthquake

adjacent to the Top Station. This was done to allow for the initial estimation of a lower range for a modulus reduction curve for the waste material. If this peak is used, the following values of $(v_s)_{avg}$, γ_{unf} and G/G_{max} are obtained for the profile adjacent to the Top Station during the Northridge earthquake:

$(v_s)_{avg} \approx 260$ m/s (850 ft/s) $\gamma_{unf} \approx 0.105 \times 0.57 \approx 0.06\%$ and $G/G_{max} \approx 0.51$

The value of $G/G_{max} \approx 0.51$ in the profile adjacent to the Top Station at a uniform strain level of 0.06% is comparable to the upper range for sands published by Seed and Idriss (1970) and used by Idriss (1990) for sands in back-calculating the motions recorded during the Loma Prieta earthquake in the San Francisco Bay area.

The motions recorded during the Northridge aftershock were not examined for this purpose.

The above considerations provided a convenient means for establishing an initial estimate of the modulus reduction curve for the waste material at this site. Obviously, the procedure outlined above is too crude for use in selecting final parameters; detailed seismic response analyses are required for final derivation of the appropriate modulus reduction and damping curves for the waste material at the OII Landfill. The remaining parts of this paper present the results of these seismic response analyses.

Analytical Procedures

The response of the landfill was modeled using the program QUAD4M (Hudson et al. 1994), which is a two-dimensional finite element code for evaluating the seismic response of soil structures. The program operates in the time domain, incorporates a compliant base, and utilizes the equivalent-linear procedure (Idriss and Seed 1968) to define dynamic soil properties. QUAD4M was used to calculate the response of the landfill during the four recorded earthquake events discussed above.

In addition, the response of the landfill was calculated by a one-dimensional equivalent linear solution using the computer program SHAKE91.

Model Characterization

Analysis Cross-Section: As noted above, an east-west cross-section was developed for a line in close proximity to both seismic stations. The location of this cross-section is shown in Fig. 1 and the cross section is shown in Fig. 2; this cross-section was used to construct the two-dimensional finite element mesh required for the seismic response analyses. This mesh is also illustrated in Fig. 2 together with the projected locations of the two seismic stations. It is noted that details of the subsurface stratigraphy were incorporated about 90 m (300 ft) to the west of the top seismic station and a similar extent to the east of seismic station at the toe. At these points the finite element mesh was extended with a constant subsurface profile to simulate free field conditions and to minimize the effects of any wave reflections from the side boundaries.

Properties of the Soil and Waste

The total unit weights and the maximum shear wave velocities used for the landfill material, the soil cover, the fill and the natural formation underlying the landfill are summarized earlier in this paper. The other key material property needed for a response calculation involves the choice of a modulus reduction curve representing the variations of G/G_{max} as a function of shear strain (G represents the strain-compatible shear modulus and G_{max} is the maximum shear modulus obtained from the measured shear wave velocity). Material damping as a function of shear strain is also needed.

Modulus Reduction and Damping Relationships: Based on the results of the field investigations and the sample descriptions, it was judged that the subsurface conditions at this landfill can be described by four different materials for the purpose of the response calculations; these four materials were clay with moderate plasticity (PI about 20 to 40), sand, waste, and the rock half-space.

The soil cover above and within the waste material and all the natural formational material were assigned modulus reduction and damping curves previously used for clays with moderate plasticity. The fill below the Toe Station was assigned modulus and damping curves previously used for sands. The rock half-space was assigned properties previously used for such material. Note that the curves for these three materials are included in the user's manual for the program QUAD4M.

The development of modulus reduction and damping curves for the waste in the landfill was the principal objective of this study.

The considerations outlined earlier in the paper, for obtaining a first order estimate of the strain levels and the possible amount of modulus reduction that may have occurred during each earthquake, were used in establishing initial estimates for the variations of G/G_{max} as a function of shear strain. In addition, wide ranges of these variations were examined before finalizing the curves finally selected as the "curves derived in this study".

A series of trial analyses were conducted shortly after the occurrence of the Northridge earthquake starting with assumed maximum shear wave velocities and considering the waste to have the modulus reduction and damping curves assigned to sand, those assigned to clay, or those suggested by Singh and Murphy (1990) in addition to other relationships that allowed for far less modulus reduction than these relationships. Because of the fact that comparisons were being made for four different events (which had varying frequency and amplitude characteristics), no one relationship was providing excellent matches with all events. In fact, the relationship that provided the best correlation between observed and calculated response at the top of the landfill was based on using a modulus reduction curve similar to that obtained for clays with high plasticity and damping comparable to that assigned to sand or clay. These were augmented by considerations similar to those outlined above which suggested very little modulus reduction in the waste/soil profile adjacent to the Top Station. Note that such comparisons were made using the maximum shear wave velocities assumed at that time for the waste.

As noted above, the maximum shear wave velocities were measured in late 1994 and the values assumed earlier proved to be closer to the lower range of the measured velocities. Upon repeating the analyses with the best estimate measured maximum shear wave velocities, the modulus reduction and damping values listed below provided the best overall comparison for all four earthquakes.

Shear Strain - percent	*G/Gmax*	Damping - percent
0.0001	1	2.5
0.0003	0.986	3.3
0.001	0.956	4.7
0.003	0.916	6.2
0.01	0.858	8.0
0.03	0.794	10.1
0.1	0.704	13.2
0.3	0.617	17.5
1	0.494	22.0

The above modulus reduction values together with those used for the clay and those used for the sand are shown in Fig. 9. The corresponding damping values are shown in Fig. 10.

Rock Outcrop Motion: The rock outcrop motion needed for two-dimensional (as well as one-dimensional) analyses was obtained by deconvolving the motion recorded at the Toe station. The deconvolution analysis was done using the one-dimensional site response program SHAKE91 (Idriss and Sun 1992) which is a slightly modified version of the original program SHAKE (Schnabel et al. 1972). The subsurface conditions and the material properties described above were used in this analysis. This process was completed for each of the four events under consideration.

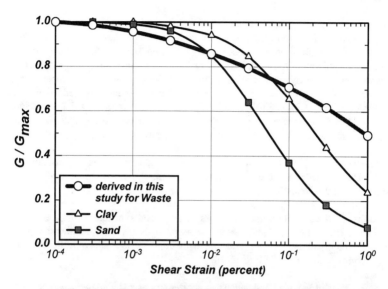

Fig. 9 Modulus Reduction Curves Used in Response Calculations

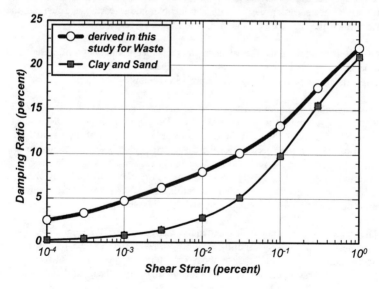

Fig. 10 Damping Curves Used in Response Calculations

Seismic Response Analyses

Seismic response analyses were completed using the finite element mesh shown in Fig. 2 and the four sets of deconvolved earthquake input motions using the best estimate maximum shear wave velocities (Figs. 3 and 4) and the modulus reduction and damping relationships described above for the four materials found at this site. In addition, the analyses were repeated using the lower range and the upper range maximum shear wave velocities with no change in the modulus reduction and damping relationships. Finally, the analyses were also repeated using the same three sets of maximum shear wave velocities and the modulus reduction and damping relationships suggested by Singh and Murphy (1990).

The results of these analyses are presented below, and comparisons are made of the calculated and recorded accelerograms and the corresponding response spectra (5% damping) at the Top Station.

Results of Two-dimensional Analyses: The results of the two-dimensional response analyses using the best estimate maximum shear wave velocities shown in Figs. 3 and 4 and the modulus reduction and damping curves shown in Figs. 9 and 10 are presented in Figs. 11 through 14.

Comparisons of recorded and calculated accelerograms and associated acceleration response spectra are presented in Fig. 11 for the Pasadena earthquake and in Fig. 12 for the Landers earthquake.   Those for the Northridge earthquake and for the Northridge aftershock are presented in Figs. 13 and 14, respectively.

Overall, it was found that the calculated and recorded ground motions, in terms of accelerograms and spectral ordinates, compare reasonably well.   Specific comparisons for each earthquake are summarized below.

(a) Pasadena Earthquake: The results for the Pasadena earthquake are shown in Fig. 11, which indicates that calculated accelerogram and its spectrum compare reasonably well with those for the motion recorded at the top of the landfill. The shape of the response spectrum for the calculated motion is similar to that for the recorded motion. Low period (high frequency) response was overpredicted; this overprediction was particularly large at a period of about 0.16 s. However, the spectral values calculated for periods longer than about 0.3 s, are almost identical to those obtained from the recorded motions during this earthquake.

The maximum shear strains calculated in the waste portion of the landfill (see Fig. 2 for extent of waste in the landfill), varied from about 0.02 to 0.05 percent in the upper part of the landfill (depth of about 7 to 30 m) and from 0.006 to about 0.02 percent in the lower portions (depth of about 44 to 99 m) of the landfill. Using a

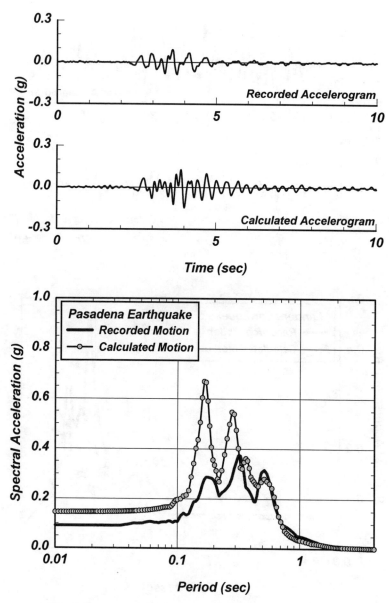

Fig. 11 Comparison of Recorded and Calculated Accelerograms
and Associated Spectra -- Pasadena Earthquake

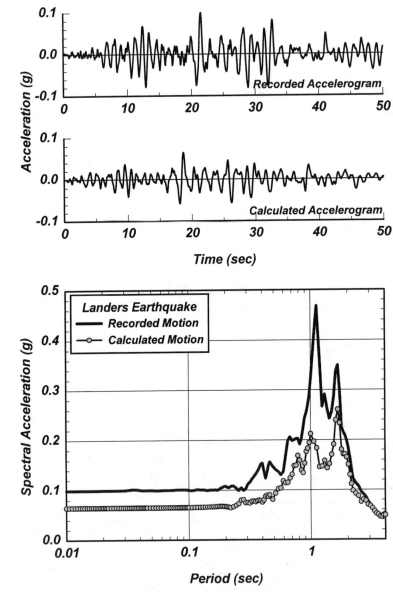

Fig. 12 Comparison of Recorded and Calculated Accelerograms
and Associated Spectra -- Landers Earthquake

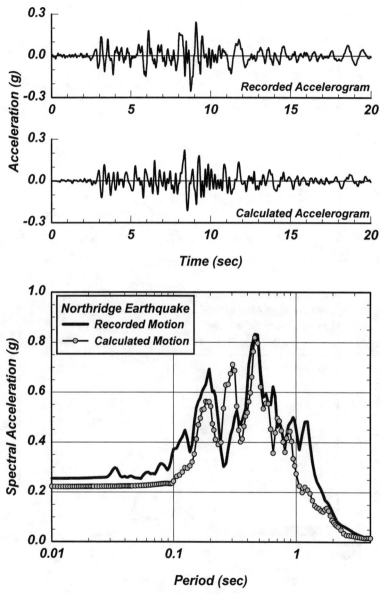

Fig. 13 Comparison of Recorded and Calculated Accelerograms
and Associated Spectra -- Northridge Earthquake

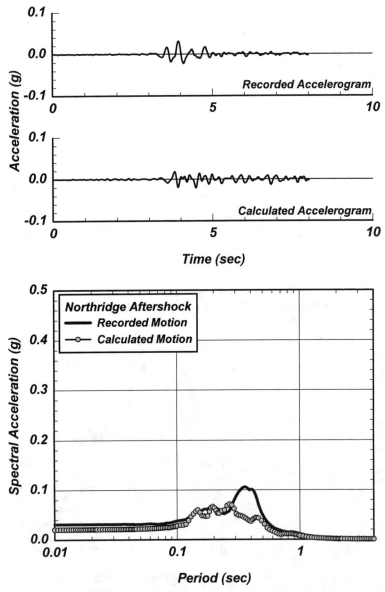

Fig. 14 Comparison of Recorded and Calculated Accelerograms
and Associated Spectra -- Northridge Aftershock

factor of 0.4 to convert maximum shear strain to equivalent uniform strain, indicates that the equivalent uniform shear strains in the waste ranged from about 0.002 to 0.02 percent, which is of the same order of magnitude as the 7×10^{-3} percent strain obtained using the peak particle velocity recorded during this earthquake and the average shear wave velocity determined from the Fourier amplitude plot, as discussed earlier in this paper.

(b) Landers Earthquake: The results for the Landers earthquake are presented in Fig. 12. The shape of the response spectrum corresponding to the calculated ground motion at the top of the landfill is similar to that for the recorded motion. It is noted, however, that the peak spectral ordinate for the recorded motion was not obtained for the computed motion. In addition, the calculated peak acceleration was only about two-thirds of the measured value.

The maximum shear strains calculated in the waste portion of the landfill, during this earthquake, varied from about 0.02 to 0.025 percent in the upper part of the landfill and from 0.01 to about 0.02 percent in the lower portions of the landfill. Using for a factor of 0.63 to convert maximum shear strain to equivalent uniform strain, indicates that the equivalent uniform shear strains in the waste ranged from about 0.006 to 0.016 percent, which somewhat smaller (albeit, of the same order of magnitude) than the 4×10^{-2} percent strain obtained using the peak particle velocity recorded during this earthquake and the average measured shear wave velocity.

(c) Northridge Earthquake: The results for the Northridge earthquake are presented in Fig. 13. The calculated accelerogram appears to be almost identical to the accelerogram of the motion recorded at the Top Station during this earthquake. The response spectrum for the calculated motion appears to match the spectrum for the recorded motion quite well both in shape and in amplitude. It is noted, however, that a small spectral peak at a period of about 0.3 s was obtained for the calculated motion and not for the recorded motion; in addition, the analysis did not capture well the spectral response for periods between 1 and 2 seconds.

The maximum shear strain calculated in the waste portion of the landfill, during this earthquake, was about 0.08 percent in the upper part of the landfill and varied from about 0.02 to 0.05 percent in the lower portions of the landfill. Using a factor of 0.57 to convert maximum shear strain to equivalent uniform strain, indicates that the equivalent uniform shear strains in the waste ranged from about 0.01 to 0.05 percent, which is comparable to the 4.6×10^{-2} percent strain obtained using the peak particle velocity recorded during this earthquake and the average measured shear wave velocity.

(d) Northridge Aftershock: The results for the Northridge aftershock, which are shown in Fig. 14, indicate similar overall comparison as those for the other events.

Overall, the Northridge and Pasadena earthquakes were the most significant events analyzed in this study because they produced the largest levels of shaking recorded at the OII Landfill. The ground motion at the top of the landfill was reasonably well predicted for the Northridge earthquake and over-predicted (for short periods) for the Pasadena earthquake.

It is useful to note that the Pasadena earthquake occurred in 1988, which is about six years prior to the implementation of the field investigation program to determine the landfill material properties. Field monitoring at the landfill, during the six years since the Pasadena earthquake, indicated that the crest of the landfill in the area of the top seismic station settled at a rate of about 30 cm (1 ft) per year (Mundy et al. 1995). This would suggest that the material properties of the waste changed between 1988 and 1994. Furthermore, it is likely that total unit weights and shear wave velocities of the landfill waste were lower in 1988 than those measured in 1994, especially for shallower depths within the landfill.

The results of the calculations using the lower range shear wave velocities are presented in Fig. 15 for the four events under consideration. As can be seen, the spectral values for the calculated motions are closer to those for the recorded motions than was obtained using the best estimate shear wave velocities (Figs. 11 through 14). The greatest improvement is obtained for the Pasadena earthquake, possibly because the lower shear velocities are more appropriate for the conditions at the landfill in 1988.

This observation may sound contradictory if the average shear wave velocity determined from the Fourier amplitude plot, as discussed earlier in this paper, represents an accurate value for the waste/soil profile adjacent to the Top Station at the time of the Pasadena earthquake. Determinations based on the use of the Fourier amplitude, however, should be used as order of magnitude estimates and should be superseded by actual measurements and other physically relevant considerations.

Results of One-dimensional Analyses: The response of the landfill was also calculated using the one-dimensional site response program SHAKE91. The one-dimensional model was developed using thickness, total unit weight, and shear modulus data specified in the two-dimensional analyses. Again, all four earthquakes were examined. The best estimate shear wave velocity profile (Figs. 3 and 4) and the modulus reduction and damping curves shown in Figs. 9 and 10 were used in these analyses.

Results of these analyses are summarized in Fig. 16. These results indicate that the

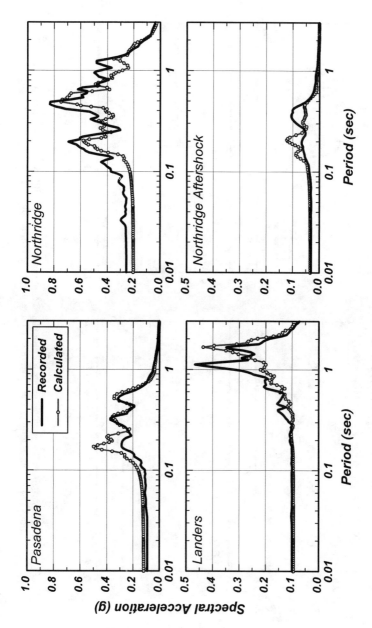

Fig. 15 Comparison of Spectral Ordinates for Calculated Motions at Top of Landfill with those for Recorded Motions Using Lower Range Shear Wave Velocities

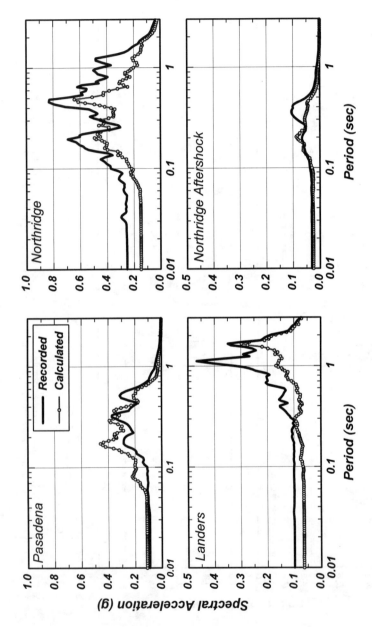

Fig. 16 Comparison of Spectral Ordinates for Calculated Motions at Top of Landfill with those for Recorded Motions Using Best Estimate Shear Wave Velocities and a One Dimensional Calculation Procedure

one-dimensional procedure can provide a reasonable estimate of the ground response at a site such as that at OII. However, the results of the two-dimensional analyses provided better correlation with the recorded values. In particular, the results for the Northridge main shock were quite poor in comparison to the results of the two-dimensional analyses.

Results Using Other Modulus Reduction and Damping Relationships for Waste

Over the past several years, investigators have utilized different modulus reduction and damping relationships for landfill waste. For example, Singh and Murphy (1990) suggested modulus reduction and damping curves for waste that were intermediate between those of the clay and peat that had been published by Seed and Idriss (1970). These curves, were used by Bray et al. (1995) in one-dimensional analyses conducted for several different landfill profiles. Recently, Kavazanjian and Matasovic (1995) developed a set of waste curves from one-dimensional back-analyses conducted for the OII Landfill. The Singh and Murphy and the Kavazanjian and Matasovic modulus reduction and damping curves are presented in Fig. 17 together with those derived in this study.

Two-dimensional seismic response analyses were also completed for the OII landfill using the modulus reduction and damping curves suggested by Singh and Murphy (1990). All four earthquakes were examined and the results are presented in Fig. 18. As can be seen in this figure, the results show excellent correlation for the Pasadena earthquake, poor correlation for the Northridge and Landers earthquakes, and reasonable correlation for the Northridge aftershock.

Seismically-Induced Permanent Lateral Deformations

Seismic considerations for a landfill are most concerned with the potential for major movements during or following an earthquake. Thus, the lateral (and possibly the vertical) inertial forces that may have the tendency to induce lateral movements in the landfill are particularly important in evaluating the potential performance of a landfill. The ratio of such a force that may act on a wedge (or segment) of the landfill divided by the mass of that wedge is defined as the seismic coefficient. Currently used simplified procedures involve integrating twice, over the entire duration of shaking, the values of such a seismic coefficient (usually designated as k and its maximum value as k_{max}) that exceed a yield coefficient (typically designated k_y) to estimate the amount of deformation of that wedge. Thus, the calculation of the seismic coefficient as a function of time for specific wedges within the landfill is a critical part of the evaluation of the seismic performance of the landfill.

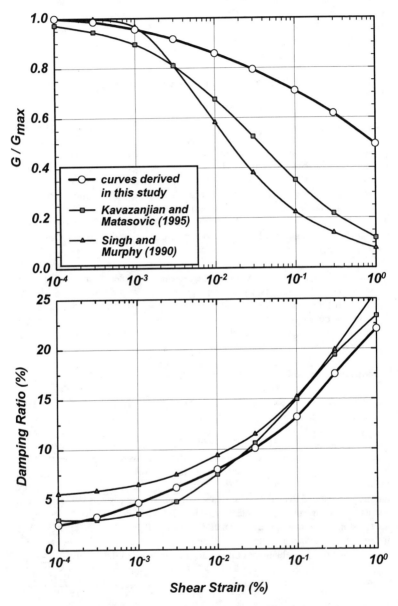

Fig. 17 Modulus Reduction and Damping Curves Used for Landfill Waste

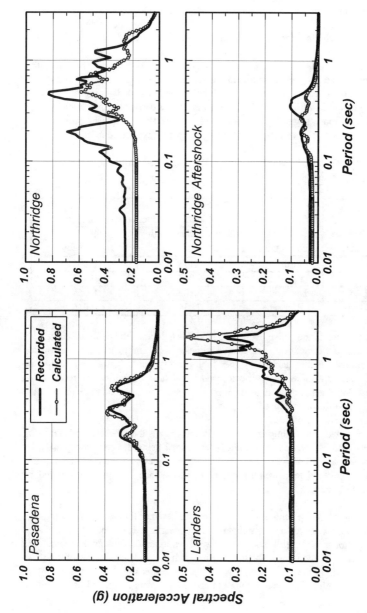

Fig. 18 Results at Top of Landfill Using a Two-Dimensional Solution, Best Estimate Shear Wave
Velocities and the Modulus Reduction and Damping Curves Suggested by Singh & Murphy (1990) for Waste

Seismic coefficients were computed for two selected wedges in the OII Landfill cross-section. One wedge extended to a depth of about 30 m (100 ft) below the surface and to a distance of about 30 m (100 ft) behind the upper crest of the landfill; the second wedge was assumed to extend to a depth of about 53 m (175 ft).

The purpose of the seismic coefficient analyses was to examine the effect that different modulus reduction and damping curves assigned to the waste would have on the results. Thus, two sets of modulus reduction and damping curves were utilized for the landfill waste; the curves were those derived in this study (Figs. 9 and 10) and those suggested by Singh and Murphy (Fig. 17). The best estimate maximum shear wave velocity profile for the landfill was utilized in these analyses.

The results of the analyses using each set of modulus reduction and damping curves are summarized below:

Wedge	Earthquake	Peak Seismic Coefficient, k_{max}, using the Modulus Reduction and Damping	
		Curves Derived in this Study	Curves Suggested by Singh & Murphy
Shallow	Northridge	0.153	0.096
Shallow	Pasadena	0.057	0.042
Deep	Northridge	0.102	0.054
Deep	Pasadena	0.034	0.026

As can be noted in the above listing, the modulus reduction and damping curves can have significant effect on the calculated values of k_{max} and hence on the estimated deformations of the landfill. The time history, and hence the integrated deformation, is also seriously affected and could further lead to lower calculated deformations when obtained using modulus reduction and damping curves that allow for significant amount of modulus reduction.

It may be noted that slightly larger seismic coefficients than those calculated using the Singh and Murphy curves would be calculated using the curves suggested by Kavazanjian and Matasovic (1995). The latter curves (see Fig. 17) incorporate lower damping and less overall modulus reduction than the curves suggested by Singh and Murphy.

Assessment of the potential for movement of a landfill due to an earthquake requires knowledge of the available shear strength, s_u, of the landfill waste. Typically, for most landfill low values have been assumed for such strength. The curve derived for modulus reduction together with the measured values of the shear wave velocities in the waste offer a way to estimate this strength for the landfill at OII. Extrapolating the curve in Fig. 9 to about 10 percent strain and using a value of G/G_{max} ranging from 0.025 to 0.035 at this strain level would result in shear strength (expressed in terms s_u/p') for the waste ranging from about $s_u/p' = 0.7$ to 1.4. These values of s_u/p' were based on calculating a vertical effective pressure, p', at the depth at which each individual maximum shear wave measurement had been made in the waste material using the total unit weights discussed earlier. Note that p' was considered to be equal to the total vertical pressure to obtain conservative values of the ratio s_u/p'. Expressing the ratio s_u/p' as an equivalent friction angle would suggest that the waste could be considered to have a friction angle ranging from about 35 to over 45 degrees, suggesting that the typical values of 30 degrees or less are highly conservative.

At this time, these values of friction angle should be considered very tentative requiring further studies and evaluations. Nevertheless, much of the excellent performance of landfills in the Northridge earthquake (as summarized by Matasovic et al, 1995) is probably due to the fact that the waste has much higher strength than had been assumed. This aspect of the problem, and including possible effects of aging, deserves further study.

Summary and Conclusions

Back-analyses were completed for four earthquakes recorded at the Operating Industries, Inc. Landfill to estimate the dynamic properties of the landfill waste. The results of the studies summarized in this paper led to the development of the modulus reduction and damping curves shown in Figs. 9 and 10. It is hoped that these curves will offer improved means for evaluating the seismic response of landfills, not only at OII, but at other landfills having similar waste

The following conclusions can be offered at this time based on the results of the studies summarized in this paper:

1.   Ground motions recorded at the landfill during four earthquakes were reasonably well matched using two-dimensional analyses and the derived modulus reduction and damping relationships for waste.

2.   The modulus reduction curve for waste derived in this study indicates less modulus reduction in the large shear strain range than modulus reduction curves for waste suggested by other investigators.

3.   The results of the two-dimensional analyses completed for the Pasadena earthquake provided closer comparison with the recorded values when the lower range of the recently measured shear wave velocities were utilized for the landfill waste. This may suggest that the properties of landfill waste material at the OII Landfill improved over time.

4.   The results of the one-dimensional analyses did not provide as good a match as those obtained using the two-dimensional analyses.

5.   The results of two-dimensional analyses conducted using waste curves suggested by Singh and Murphy (1990) indicated that the landfill responded in a softer manner than observed in the measured recordings.

6.   Seismic coefficients computed for selected wedges at the landfill, using the modulus reduction and damping curves developed in this study for waste, are significantly higher than those computed using the waste curves suggested by others, such as the curves by Singh and Murphy (1990) or the curves by Kavazanjian and Matasovic (1995). Care, however, should be exercised to insure the use of reasonable values for the shear strength of the material comprising the wedge under consideration.

7.   The use of the recorded motions to obtain first order approximations of the characteristics of the waste/soil profile adjacent to the Top Station proved to be a value tool. The results, however, should be superseded by detailed seismic response calculations for use in assessing the potential performance of the landfill.

The motions recorded at the OII Landfill have proven to be of paramount usefulness. Many other landfills have initiated the installation of similar seismic stations. The recordings from future earthquakes at OII and at the other landfills should provide a wealth of information for use in future applications and for improving the definition of material properties and the methods of analysis.

Acknowledgments

The study reported in this paper was supported by CDM Federal Programs Corporation, Under Contract to the US Army Corps of Engineers for the US Environmental Protection Agency (USEPA), and by the Center for Geotechnical Modeling at the University of California at Davis. The participation of the senior author in this study was supported in full by the Center for Geotechnical Modeling. The analyses were completed at CDM Federal Programs Corporation. Mr. Ali Bastani, formerly with CDM Federal Programs Corporation, assisted in the initial analysis. Mr. Michael Schwan, Manager of the Walnut Creek Office of CDM Federal Programs Corporation, was instrumental in making the facilities at his office available for this study. Mr. Jeff Di Donato drafted some of the figures included in this paper. The authors are grateful to these organizations and individuals for their support.

The work leading to the preparation of the material presented in this paper was funded, in part, by the USEPA. The preparation of the paper, however, was not funded by the USEPA and has not been subject to the Agency's review. Therefore, the contents of this paper do not necessarily reflect the Agency's views and no official endorsement is implied or should be inferred.

References

Bray, J. D., Augello, A. J., Leonards, G A, Repetto, P.C., and Byrne, R. J. (1995) "Seismic Stability Procedures for Solid-Waste Landfills" Journal of Geotechnical Engineering, ASCE, Vol. 121, No. 2, February.

Hudson, M., Idriss, I. M., and Beikae, M. (1994) "QUAD4M: A Computer Program to Evaluate the Seismic Response of Soil Structures using Finite Element Procedures and Incorporating a Compliant Base" Center for Geotechnical Modeling, Department of Civil and Environmental Engineering, University of California, Davis, May.

Idriss, I. M. and Seed H. B. (1968) "Seismic Response of Horizontal Soil Layers" Journal of the Soil Mechanics and Foundations Division, ASCE, Vol. 94, No. SM4, July.

Idriss, I. M. (1990) "Response of Soft Soil Sites during Earthquakes" Proceedings, H. Bolton Seed Memorial Symposium, Volume 2, BiTech Publishers, Ltd., Vancouver, B. C., May.

Idriss, I. M. and Sun, J. I. (1992) "SHAKE91 - A Computer Program for Conducting Equivalent-Linear Seismic Response Analyses of Horizontally Layered Soil Deposits" Program modified based on the original program SHAKE; Center for Geotechnical Modeling, Department of Civil and Environmental Engineering, University of California, Davis.

Kavazanjian, E. and Matasovic, N. (1995) "Seismic Analysis of Solid Waste Landfills" Proceedings of Geoenvironment 2000, Volume 2, Geotechnical Special Publication No. 46, ASCE, 1066 - 1081.

Matasovic, N, Kavazanjian, E., Augello, A. J., Bray, J. D., and Seed, R. B. (1995) "Solid Waste Landfill Damage Caused by 17 January 1994 Northridge Earthquake," in Woods, Mary C. and Seiple, W. Ray, eds., The Northridge, California, Earthquake of 17 January 1994; California Department of Conservation, Division of Mines and Geology Special Publication, 116 p.

Mundy, P. K., Nyznyk, J. P., Bastani, S. A., Brick, W. D., Clark, J., and Herzig, R. (1995) "Geotechnical Monitoring of the OII Landfill" Proceedings Paper, ASCE National Convention, San Diego.

Newmark, N. M. (1965) "Effects of Earthquakes on Dams and Embankments" Geotechnique, Vol. 15, No. 2, June.

Schnabel, P. B., Lysmer, J., and Seed, H. B. (1972) "SHAKE: A Computer Program for Earthquake Response Analysis of Horizontally Layered Sites" Report No. UCB/EERC-72/12, Earthquake Engineering Research Center, University of California, Berkeley, December.

Seed, H. B. and Idriss, I. M. (1970) "Soil Moduli and Damping Factors for Dynamic Response Analysis" Report No. UCB/EERC-70/10, Earthquake Engineering Research Center, University of California, Berkeley, December.

Singh, S. and Murphy, B. (1990) "Evaluation of the Stability of Sanitary Landfills" Geotechnics of Waste Fills - Theory and Practice, ASTM STP 1070, Arvid Landva, G. David Knowles, eds., American Society for Testing and Materials, Philadelphia, 240 - 258.

Trifunac, M. D. and Brady, A. G. (1975), "A Study of the Duration of Strong Earthquake Ground Motion", Bulletin of the Seismological Society of America, Vol. 65, pp 581-626.

U. S. Environmental Protection Agency (1994) "Landfill Response to Seismic Events, Evaluation of Earthquake Ground Motions Recorded at Operating Industries, Inc. Landfill Superfund Site in Monterey Park, California during the January 17, 1994 Northridge Earthquake" Prepared for the USEPA by Hushmand Associates, April.

U. S. Environmental Protection Agency (1995) "Processing and Spectral Analysis of Eighteen Earthquakes Recorded at the Operating Industries, Inc. Landfill in Monterey Park, California - Final Report" Prepared for the USEPA by Pacific Engineering and Analysis, March.

HAZARD ANALYSIS FOR A LARGE REGIONAL LANDFILL

by Edward Kavazanjian, Jr.[1] M. ASCE, Rudolph Bonaparte[2] M. ASCE,
Gary W. Johnson[3] M. ASCE, Geoffrey R. Martin[4] M. ASCE, and
Neven Matasović[1] A.M. ASCE

ABSTRACT

The Eagle Mountain landfill, with a proposed airspace of 510 million cubic meters, will be upon completion one of the largest landfills in the world. Solid waste slopes at the landfill will be up to 350 m in height from toe to crest and the thickness of the solid waste will exceed 200 m. The unprecedented size of the landfill, its seismic exposure, and uncertainty over implementation of new federal regulations made preliminary design and permitting of the facility a challenge. In addressing this challenge, the seismic hazard analysis for the landfill was done using two different approaches: the prescriptive approach contained in Federal regulations and the alternative approach contained in California regulations. Application of the results of these seismic hazard analyses to seismic performance assessment of the landfill demonstrates that neither approach can be unconditionally described as more or less stringent than the other. Both approaches are considered to form the basis for design of a landfill with a high degree of seismic resistance that provides a high level of protection to the environment.

---

[1]GeoSyntec Consultants, 16541 Gothard Street, Suite 211, Huntington Beach, CA 92647

[2]GeoSyntec Consultants, 1100 Lake Hearn Drive N.E., Suite 200, Atlanta, GA 30342

[3]Mine Reclamation Corporation, 960 Tahquitz Canyon Way, Suite 204, Palm Springs, CA 92262

[4]Department of Civil Engineering, University of Southern California, Los Angeles, CA 90089

INTRODUCTION

Design analyses for environmental permitting of the Mine Reclamation Corporation's Eagle Mountain Landfill and Recycling Center, located in Riverside County, California, took place primarily between fall 1991 and fall 1993. The final version of United States Environmental Protection Agency (USEPA) regulations for seismic design of Municipal Solid Waste Landfill Facilities (MSWLF), contained in Subtitle D of the Resource Conservation and Recovery Act (Subtitle D), were published in October 1991 and became effective in October 1993. Therefore, there was little to no precedent to draw from with respect to interpretation of the Subtitle D regulations. California regulations for seismic design of MSWLF pre-date Subtitle D and their interpretation was relatively well established. Because of uncertainty over the interpretation and implementation of the new federal regulations and because of the owner's desire to design the facility to comply with all applicable standards, analyses were conducted to evaluate compliance of the proposed Eagle Mountain facility with both the prescriptive Subtitle D and alternative California standards for seismic design of MSWLF.

The federal Subtitle D regulations for seismic design of MSWLF provide a prescriptive seismic design standard based upon the maximum horizontal acceleration (MHA) in lithified earth (bedrock) at the site with a 90 percent or greater probability of not being exceeded in 250 years, as depicted on a seismic hazard map. The MHA is typically evaluated from United States Geological Survey (USGS) Map Sheet MF-2120 (USEPA, 1992). This map is commonly referred to as the "Algermissen" map. As an alternative to evaluating the MHA from a seismic hazard map, the Subtitle D regulations also provide for the alternative approach of evaluating the MHA based upon a site-specific analysis. However, USEPA provides little guidance on what constitutes an acceptable site-specific analysis (USEPA, 1995). California regulations specify that MSWLF be designed to withstand the Maximum Probable Earthquake (MPE) in accordance with California Department of Natural Resources Division of Mines and Geology (CDMG) guidelines. CDMG (1975) defines the MPE as the *maximum* earthquake expected to impact the site in a 100 year period. Under CDMG guidelines, the MPE is evaluated in a deterministic manner from a site-specific seismic hazard analysis.

There are a number of differences between the prescriptive federal standard and the California standard for the design earthquake for MSWLF. The prescriptive Subtitle D seismic design criteria explicitly requires that the earthquake with the maximum horizontal acceleration be used in design. However, the event with the maximum horizontal acceleration is not necessarily the event with the greatest damage potential. A larger magnitude earthquake at a greater distance than the event generating the MHA may be associated with a lower peak horizontal ground acceleration (PGA) than the MHA but may have a greater damage potential. One advantage of the California regulations is that the *maximum* earthquake is commonly interpreted as the earthquake with the greatest damage potential, and not simply as the earthquake with the largest PGA. Thus, under California regulations both near field events generating relatively high PGAs and far field events generating lower PGAs are considered in determining the design earthquake. Furthermore, the magnitude of the California MPE design event is explicitly evaluated while the Algermissen map, and most other published seismic hazard maps, provide no direct means of assessing the magnitude associated with the MHA. Another advantage of the California regulations is that the design earthquake is evaluated from a site-specific hazard analysis rather than from a published map. This facilitates incorporation of recent developments in understanding of regional seismology, tectonics, and earthquake ground motion propagation that may not be reflected in published ground motion maps.

In California's application to USEPA for approval of its regulatory program as in compliance with Subtitle D, there was no mention of any proposed changes to the state regulations for seismic design. At the time the design and permitting activities for the Eagle Mountain facility were proceeding, the California application had not been approved by USEPA and there was uncertainty as to the outcome of the approval process. Furthermore, the Mine Reclamation Corporation (MRC), the owner of the facility, desired that the Eagle Mountain facility be designed to conform to all applicable standards. Therefore, the Eagle Mountain Landfill was designed to conform to both the prescriptive Subtitle D standard and the California standard for seismic design.

In accordance with the prescribed federal standard, the MHA was evaluated as 0.56 g by interpolation from the Algermissen map. A site-specific probabilistic analysis was performed to validate this MHA value and to evaluate the distribution of earthquake magnitude and site-to-source distance associated with the MHA. For design purposes, a representative magnitude of between 6.0 and 6.5 and a site-to-source distance of approximately 8 km were established for the MHA event based upon this distribution. In accordance with the California standard, the MPE was established as a magnitude 7.7 event on the San Andreas fault at a distance of 53 km from the site. The free field PGA in bedrock at the site from the MPE was established to be 0.14 g. Suites of representative acceleration time histories from the catalog of available time histories were assigned to both the MHA and MPE design events. Seismic performance analyses were conducted using both the MHA and MPE design events to demonstrate that the proposed design complied with both the prescriptive federal and alternative state regulatory standards.

PROJECT DESCRIPTION

MRC intends to develop, own, and operate the Eagle Mountain Landfill as a California Class III nonhazardous solid waste disposal facility. The project site, shown in Figure 1, located about 270 km east of Los Angeles, in northeastern Riverside County, California, is the site of a former open pit iron mine and ore

Figure 1.   Site Location Map

processing facility. The proposed landfill will occupy approximately 750 ha. The landfill will be a canyon fill type of facility with waste slopes that measure over 350 m from toe to crest and waste depths of over 200 m in some locations. Nonhazardous solid waste will be transported to the landfill, primarily by rail from transfer stations located in southern California and in lessor quantities by truck from local sources. At the ultimate design capacity of 18,000 tonnes of solid waste per day, the total available airspace for the landfill of 510 million cubic meters will be filled up in 78 years.

The waste containment system for the Eagle Mountain Landfill, shown schematically in Figure 2, was designed in accordance with Subtitle D regulations. The liner system for the landfill will consist of a composite liner overlain by a leachate collection and removal system (LCRS). In the base areas of the landfill (i.e., the bottom areas of canyons and pits having slopes 3H:1V (horizontal:vertical or flatter), the composite liner will consist of a high density polyethylene (HDPE) geomembrane liner with a nominal thickness of 2 mm underlain by a low-permeability soil liner with a hydraulic conductivity of not more than 1×10^{-7} cm/s and a thickness of at least 0.6 m. In the side slope areas of the landfill (i.e., the areas of the landfill having slopes steeper than 3H:1V), and in the bench and ridge areas of the landfill (i.e., the areas above the base having slopes of 3H:1V or flatter) the composite liner will consist of a HDPE geomembrane liner with a nominal thickness of 2 mm underlain by a geosynthetic-clay liner (GCL) with a hydraulic conductivity of not more than 5×10^{-9} cm/s.

The LCRS will be composed of a granular blanket drain (drainage layer) with a minimum slope of 4 percent. In the base areas of the landfill and in bench and ridge areas, the drainage layer will consist of a 0.45-m thick layer of coarse gravel with a hydraulic conductivity of at least 1.0 cm/s. In the side slope areas of the landfill, the LCRS will consist of a single layer of coarse sand and gravel at least 0.9 m thick with a hydraulic conductivity of at least 1×10^{-2} cm/s.

LCRS drainage corridors will be constructed at the low points of major graded canyon areas. The drainage corridors will be constructed using coarse gravel material with a hydraulic conductivity of at least 1×10 cm/s. Liquids in the LCRS drainage layer will flow to the drainage corridors which, in turn, will convey the liquids to one of ten LCRS sumps at the lowest points within the landfill footprint. Liquid that drains into the LCRS sumps will be removed by pumping using dedicated submersible pumps placed in side slope risers at the low point of the sumps. The pumps will operate when sufficient liquid is available to allow for proper pump operation. Even though only one is needed, two side slope risers will be installed in each sump to provide redundancy, additional capacity, and the ability to flush the sumps in the unlikely event it is necessary.

The final cover system for the Eagle Mountain Landfill will consist of a 1-mm thick flexible polyolefin geomembrane cap resting on a 0.6-m thick foundation

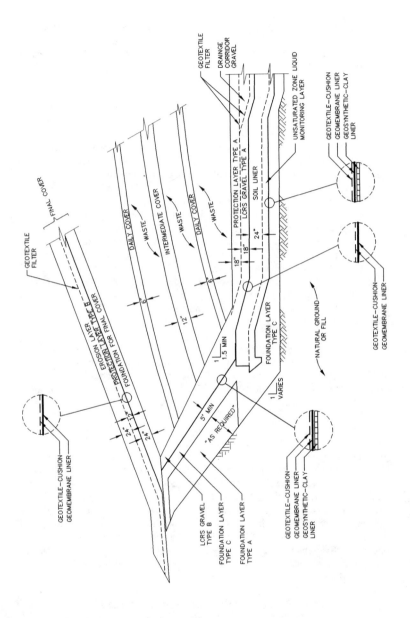

Figure 2.   Subtitle D Waste Containment System

for final cover layer. The upper 0.45 m of the foundation for final cover layer will have a hydraulic conductivity no greater than 1 x 10^{-5} cm/s. The geomembrane cap will be overlain by a nonwoven geotextile cushion layer, a 0.3-m thick soil protection layer, and a 0.6-m thick erosion layer consisting of cobble- and boulder-sized material.

An active gas extraction system will be used to remove landfill gas generated by the decomposition of waste. The active gas extraction system will be comprised of about 1,000 vertical gas extraction wells placed within the waste mass. Construction of the active gas extraction system will occur progressively as waste filling progresses. The active gas extraction system will be operated under a small vacuum (i.e., less than atmospheric pressure) which will promote gas flow from the waste mass towards the gas extraction wells.

SEISMIC SETTING

The Eagle Mountain landfill site is located along the northeastern edge of the Eagle Mountains, in an area of high topographic relief. The terrain is rugged, consisting of a series of semi-parallel southeastward sloping canyons. The Eagle Mountains extend into the Basin and Range Geologic Province at the eastern extremity of the Southern California Transverse Ranges. Bedrock within the project area consists of Paleozoic age meta-sedimentary rocks which have been intruded by Mesozoic igneous rocks.

The site is located at the eastern edge of the zone of major historic recorded seismic activity in southern California. Figure 3 shows earthquakes of magnitude 4 or greater recorded since 1800 within 320 km of the site (Earth Mechanics, 1992). The Coachella Valley segment of the San Andreas fault is located approximately 53 km west of the site. Other significant known faults with respect to the seismicity of the project site include the Pinto Mountain fault, located approximately 45 km northwest of the site, and the Blue Cut fault, located approximately 6 km to the north of the site. Closer to the site lie a number of small faults that are generally considered to be ancient features, including the Substation, Victory Pass, and Porcupine Wash faults (Schell, 1992). In addition to these fault-specific potential seismic sources, the site is subject to non-fault specific "random" seismicity associated with the Southeast Transverse Ranges seismo-tectonic zone.

Several local faults were mapped at the site as part of investigations undertaken during permitting of the landfill (Proctor, 1992; Shlemon, 1992). These investigations produced strong evidence that these faults had not undergone surface displacement during the Holocene Epoch, as required by federal and state regulations. This evidence included geologic field mapping within the mine pit and geochemical analyses that indicated that igneous dikes of Mesozoic age were not displaced by faulting and trench logging and analyses of aerial photography and soil stratigraphic studies indicating that the varnished soil surface in the adjacent desert terrain had not

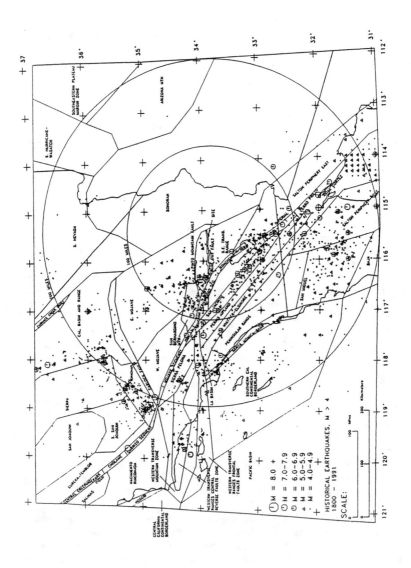

Figure 3.   Regional Seismicity (Earth Mechanics, 1992)

been disturbed for at least 40,000 to 100,000 years. In fact, the field reconnaissance and interpretation of aerial photography indicated that the last instance of active faulting at the site may have been as long as 5,000,000 years ago.

Figure 3 also shows the relationship of the Eagle Mountain Landfill site to major Quarternary fault zones in the state of California (Jennings, 1992). Major Quarternary fault zones capable of generating strong ground shaking at the site included the San Andreas fault (divided into the Coachella Valley, Mojave, Carrizo Plain, and San Bernardino segments for the purposes of this study), the Imperial fault, the San Jacinto Fault, the Pinto Mountains fault, and the Blue Cut fault. In addition to these well defined fault zones, the Southeast Transverse Ranges and Eastern Mojave, and Sonoran tectonic provinces were included as area sources (sources of diffuse, non-fault specific earthquakes) in the seismic hazard evaluation.

SEISMIC HAZARD ASSESSMENT

Prescriptive Subtitle D Hazard Assessment

The prescriptive Subtitle D design value for the MHA at the site, interpolated from the most recent Algermissen, et al. (1990) map, was slightly less than 0.6 g. Because of concerns about the accuracy of the Algermissen map and because the landfill seismic performance analyses required knowledge of the magnitude of the design event and the Algermissen map and supporting documentation provide little guidance as to the magnitude associated with the MHA value form the map, a site-specific seismic hazard analysis was performed. The site-specific probabilistic analysis was used to substantiate the Algermissen map value of the MHA and determine the associated representative magnitude. To characterize the regional and local seismicity for input to the site-specific analysis, a detailed seismo-tectonic investigation was performed (Schell, 1992). This investigation included aerial reconnaissance, aerial photograph interpretation, field geologic investigation, compilation of the earthquake history of the area, definition of the seismic sources capable of generating strong ground motions at the site, and evaluation of earthquake recurrence relationships. The investigation screened known seismic sources within approximately 320 km of the site and then focused on the significant sources identified with approximately 160 km of the site.

Figure 3 is a plot of all known historical earthquakes of magnitude greater than 4.0 within a 320 km radius of the site for the period from 1800 to 1991. This figure also shows the major faults, fault zones, and seismo-tectonic source zones. The data shown in Figure 3 are only considered complete for the past 60 years, since establishment in 1932 of the Southern California Seismic Network jointly administered by the USGS and California Institute of Technology. Epicentral locations of events recorded in this period may be considered accurate to within a few kilometers (Given et al. 1987). Prior to 1932, only events large enough and close enough to be felt in populated areas are known, and locations of these events are inferred based upon either observations of surface rupture or reports of observations of shaking intensity.

Figure 3 shows the site to lie within the Southeast Transverse Ranges seismo-tectonic zone on the eastern edge of the region of high historical seismicity in southern California. Furthermore, the site is south and east of the zones of highest historical seismic activity within the Southeast Transverse Ranges. Most seismicity within the Southeast Transverse Ranges province is associated with the San Andreas Fault Zone in the northwest corner of the zone. Much of the remaining seismicity within the zone is concentrated between the Blue Cut and Pinto Mountain faults to the north of the site. The closest recorded seismic event to the site on the historical record is an event of magnitude 4.0 to 4.9 approximately 8 km south of the site. The largest events of greater than magnitude 5 were two events of between magnitude 5.0 and 5.9 approximately 21 km northwest of the site, in the region between the Blue Cut and Pinto Mountain faults. The closest recorded event of greater than magnitude 6.0 is approximately 80 km west of the site and appears to be associated with the San Andreas Fault Zone. Based upon these magnitudes and distances, and using the attenuation relationship developed by Sadigh as reported by Joyner and Boore (1988) for rock sites in southern California, the strongest ground motion at the site from events in the historical record would be estimated as 0.15 g using mean attenuation rates, and 0.27 g using mean plus one standard deviation attenuation rates.

For the Eagle Mountain Landfill hazard analyses, activity rates for the principal causative faults, fault zones, and seismo-tectonic provinces were estimated using a combination of instrumental records of historic seismicity, detailed geologic information for faults with evidence of surface faulting, and regional geologic information on the long-term geologic slip rate of the fault or seismo-tectonic province. For each principal causative seismic source, all three types of information were compiled, evaluated for reliability (e.g., completeness of the historical catalog, level of detail of field studies), and integrated to provide an estimate of fault activity consistent with local geology and regional tectonics. Fault activity rates were then described using the Gutenberg-Richter logarithmic relationship. Complete details of this activity assessment are given by Schell (1992). Important aspects of the activity assessment are summarized below.

For the Blue Cut fault, the closest recognized active or potentially active fault to the project site, activity rates were based primarily upon field studies performed specifically for this project (Schell, 1992). For the Pinto Mountain and San Andreas faults, the other fault-specific sources among the principal causative sources contributing to seismicity at the site, field studies reported in the literature were the primary data used to evaluate activity rates. Activity rates for the seismo-tectonic province source zones were based upon a combination of historic seismicity data (i.e., for the small magnitude end of the range, where the historic catalog is considered complete) and regional geologic slip rates (i.e., for the large magnitude end of the range). Seismicity associated with fault-specific sources (e.g., Blue Cut, Pinto Mountain, and San Andreas faults) was removed from the historic data and geologic slip rates in evaluating seismo-tectonic source activity rates to eliminate "double counting" of seismic events. The remaining seismo-tectonic source activity rates include both non-fault specific random seismicity and seismicity that may be associated with known faults that have not exhibited ground surface displacement in Holocene or late Pleistocene time, including small local faults such as the Victory Pass and Substation faults.

Figure 4 shows the resulting activity rates for all seismic sources used in the seismic hazard analysis. Table 1 presents recurrence intervals for each of the primary contributing sources considered in the seismicity assessment for events of greater than magnitude 4.5 (i.e., considered to be the minimum magnitude of a damaging event) and for events within 0.5 of the maximum magnitude event considered likely for each source based upon geology and tectonics (i.e., the MCE). PGAs associated with the MCE for each source, calculated using the mean attenuation equation for rock sites in southern California developed by Sadigh as reported by Joyner and Boore (1988), are also presented in this table for information purposes.

The probabilistic seismic hazard assessment was performed by Earth Mechanics (1992) using the computer program SEISRISK III (Bender and Perkins, 1987). Figure 5 shows the results of the probabilistic hazard analysis performed using the data in Figure 4 and Table 1. Despite the difference in assumptions between the analyses used to generate the Algermissen maps and the site-specific probabilistic analyses, the site-specific probabilistic seismic hazard analysis yielded 0.56 g as the PGA with a 90 percent probability of not being exceeded in 250 years, in substantial agreement with the prescriptive Subtitle D MHA from the Algermissen map. The magnitude distribution corresponding to this prescriptive MHA is also shown in Figure 5. This magnitude distribution indicates that over 98 percent of the earthquakes contributing to this acceleration were from the Southeast Transverse Ranges regional source and were of magnitude 6.5 or smaller at a distance of 8 km or less. Based upon this result, a magnitude of between 6.0 and 6.5 at a distance of approximately 8 km was assigned to the 0.56 g PGA to describe the MHA design earthquake. This design earthquake may be attributed to both non-fault specific random seismicity in the vicinity of the site and to local non-Holocene faults such as the Substation and Victory Pass faults.

MPE HAZARD ASSESSMENT

As discussed previously, the MPE earthquake specified in California regulations was also assessed for the purposes of landfill design. Based on a review of the data in Table 1, it was concluded that in assessing the maximum (most damaging) earthquake at the site, the potential impact of a large earthquake on the

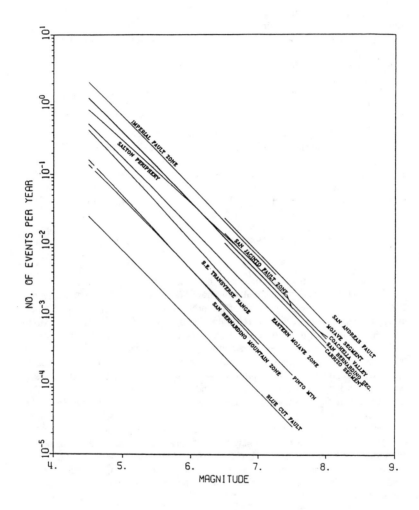

Figure 4.   Source Zone Activity Rates

Table 1. Significant Seismic Sources Within 100 km of the Eagle Mountain Landfill Site

FAULT OR FAULT ZONE	CLOSEST DISTANCE, km	LENGTH (L), km OR AREA (A), km²	MAXIMUM CREDIBLE EARTHQUAKE[1] MAGNITUDE (M_{max})	RECURRENCE INTERVAL, years		MAXIMUM CREDIBLE EARTHQUAKE PEAK HORIZONTAL ACCELERATION,[2] g's
				M ≥ 4.5	M ≥ (M_{max} - 0.5)	
Blue Cut Fault	6	L - 83	7.5	39.5	12,500	0.48
Pinto Mountain Fault	45	L - 80	7.2	7.2	2,290	0.10
Southeast Transverse Ranges Zone[3]	5[4]	A - 6,737	6.75	2.3	166	0.49
San Bernardino Mountains Zone	90	A - 2,156	7.0	6.2	778	0.03
Eastern Mojave Zone	11	A - 22,008	7.5	1.9	573	0.41
Sonoran Zone	22	A - 115,487	6.5	44.7	1,412	0.15
Salton Zone	55	A - 32,269	7.0	1.2	73.6	0.07
San Andreas Fault:[5]						
- Coachella Valley Segment	53	L - 69	8.0	69.5	695	0.16
- San Bernardino Segment	65	L - 125	8.0	0.8	795	0.14

Notes: (1) Maximum Credible Earthquake (MCE) is the "maximum earthquake that appears capable of occurring under the presently known tectonic framework" (CDMG, 1975). The MCE is presented as a means of indicating one of the relative differences among source zone source characteristics.

(2) Using mean attenuation relationship of Sadigh as reported by Joyner and Boore (1988).

(3) Includes Substation, Victory Pass, and Porcupine Wash faults.

(4) Site is within S.E. Transverse Range. Minimum distance assumed to be 5 km based upon depth of micro-seismic activity.

(5) Minimum magnitude equal to 6.5 for Coachella Valley Segment. Maximum 8.0 maximum event assumes simultaneous rupture of Coachella Valley, San Bernardino, and Eastern Mojave Segments.

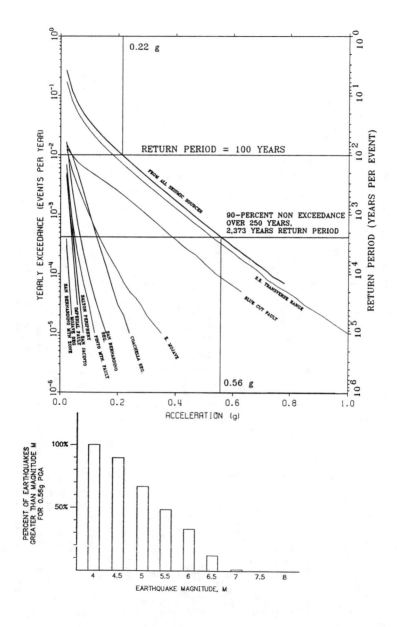

Figure 5.   Probabilistic Hazard Analysis Results

San Andreas fault should be considered. Not only is the San Andreas fault zone considered capable of generating the largest magnitude earthquake impacting the site, it is considered by many earth science professionals as having a relatively high likelihood of rupture during the 78-year life of the landfill. Recent paleoseismic studies (e.g., Sieh, 1978; Weldon and Sieh, 1985) have extended the historical record of major earthquakes back over multiple major events spanning several hundred years on the three segments of the San Andreas fault closest to the Eagle Mountain Landfill site (the Mojave, San Bernardino Mountains, and Coachella Valley segments). These studies have established the date of the last event and expected (mean) recurrence interval of events on each of these fault segments with relative certainty. For each segment, the time since the last event is close to, or greater than, the expected recurrence interval for that segment. MPE events from all of the other identified seismic sources are accounted for by either the prescriptive Subtitle D design parameters of a magnitude 6.5 event generating a PGA of 0.56 g or by the parameters for the San Andreas event.

Based upon the findings of the Working Group on the Probabilities of Future Large Earthquakes in Southern California (WGPFLESC, 1992), a magnitude 7.7 event due to concurrent rupture of the Coachella Valley and San Bernardino Mountains segments of the San Andreas fault was identified as representative of the MPE event along the southern California portion of the San Andreas fault system. Using the mean attenuation relationship of Sadigh as reported by Joyner and Boore (1988) and using the closest approach of the fault to the site (53 km) as the source to site distance, this MPE event was associated with a free field PGA of 0.14 g at the Eagle Mountain Landfill site.

For information purposes, based upon the WGPFLESC (1992) studies, a magnitude 8.0 earthquake event associated with simultaneous rupture of three segments of the San Andreas fault was identified as the Maximum Credible Earthquake (MCE) for the southern California portion of the San Andreas fault system. The MCE is defined by CDMG (1975) as the maximum earthquake capable of occurring under the presently known tectonic framework. Using the mean attenuation relationship from Sadigh, this MCE event was associated with a free field PGA of 0.16 g in bedrock at the Eagle Mountain Landfill site.

REPRESENTATIVE TIME HISTORIES

A suite of representative time histories were selected for both the prescriptive Subtitle D event and the California MPE event for use in the seismic performance analyses of the landfill. Time histories were selected to represent the "rock site" foundation conditions and the mechanism, magnitude, PGA, and source to site distance of the design events. A suite of eleven different accelerograms recorded in California from earthquakes of magnitude 5.9 to 7.1 were used to characterize the prescriptive Subtitle D design earthquake. Each of the time histories was scaled to a PGA of 0.56 g for the seismic performance analyses. These time histories were as follows:

- the Gilroy record from the magnitude 5.9 Coyote Lake earthquake; PGA of this record was 0.32 g;

- both components of the magnitude 5.9 Whittier recording at the Garvey Reservoir Control Building; PGA for these records were 0.37 g and 0.48 g;

- both components of the Cholame 5 record from the 1966 magnitude 6.0 Parkfield event; the PGA for these records were 0.35 g and 0.43 g;

- both components of the Cholame 8 record from the 1966 magnitude 6.0 Parkfield event; the PGA for these records were 0.24 g and 0.28 g;

- both components of the Corralitos record from the 1989 magnitude 7.1 Loma Prieta earthquake; the PGA for these records were 0.48 g and 0.63 g; and

- both components of the University of California, Santa Cruz record from the 1989 magnitude 7.1 Loma Prieta earthquake; the PGA for these records were 0.41 g and 0.44 g.

Selection of representative recorded accelerograms for the MPE design event was complicated by the lack of available rock site records from large magnitude earthquakes. The following four accelerograms were selected to represent the magnitude 7.7 MPE event at the site:

- both components of the Taft record from the magnitude 7.7 Kern County earthquake of 1952; the PGA for these records were 0.16 g and 0.18 g.

- the Silent Valley record from the magnitude 7.4 Landers earthquake of 1992; the PGA of this record was 0.05 g; and

- the Twenty-Nine Palms record from the magnitude 7.4 Landers earthquake of 1992; the PGA of this record was 0.06 g.

In part to compensate for the difference in magnitude between the Landers records and the MPE and in part to provide a conservative basis for design, the above records were scaled to the MCE PGA of 0.16 g for use in the seismic performance analyses of the landfill.

APPLICATION TO LANDFILL PERFORMANCE ANALYSES

Seismic performance analyses for the landfill included pseudo-static stability analyses to calculate the yield acceleration of the waste mass and cover system, seismic response analyses of the waste mass to estimate the peak acceleration and acceleration time history of the cover system and the peak average acceleration and average acceleration time history of the waste mass, and Newmark deformation analyses to calculate permanent deformations of the waste mass and cover system for cases where the yield acceleration was lower than the peak acceleration.

Figure 6 shows the variety of failure surfaces considered in the stability analyses of the waste mass. Stability analyses of the cover system considered only veneer type "infinite slope" failure surfaces along cover interfaces. Results of the stability analyses indicated that the yield acceleration varied from 0.05 g to 0.15 g for failure surfaces passing along the liner and from 0.15 g to 0.25 g for suraces along the final cover, depending on the choice of materials. In the seismic response and deformation analyses, the sensitivity of the calculated permanent deformation to yield acceleration was evaluated to establish what yield acceleration values, and hence what materials, were acceptable for landfill construction.

Since the landfill was founded upon rock, response analyses were not required to evaluate the effect of local soil conditions on the earthquake motions. However, response analyses were required to evaluate the response of the landfill mass to the seismic motions. While the thickness of the waste mass exceeded 200 m in some isolated locations, waste mass thickness in the slope areas was generally less than or equal to 90 m. Therefore, landfill response analyses were performed for waste thicknesses of 15, 45 and 90 meters.

Response analyses were performed using the computer program SHAKE (Schnabel et al., 1972). Material properties of the waste were based upon the results of field investigations and laboratory tests conducted for the Puente Hills landfill in Los Angeles (Earth Technology, 1988). With respect to the response of the waste mass, the thinnest column of waste (15 m) always showed the maximum seismic response (highest PGA, largest permanent deformations). With respect to the response of the cover system, the maximum response varied depending on the match between the natural period (first mode) of the waste column and the predominant period of the input earthquake motion. In the analysis for the magnitude 6.0 to 6.5 design earthquake of intensity 0.56 g, the average acceleration of the waste mass always showed attenuation, while the peak acceleration of the cover occasionally indicated amplification from the PGA. Amplification occurred in cases where the predominant frequency of the waste column closely matched the predominant frequency of the input acceleration. The worst case occurred with the Gilroy record for a 15 m waste thickness, where a peak acceleration of 0.88 g at the top of the cover was calculated. Even in this case, the average acceleration of the waste mass showed significant attenuation.

MODES OF POTENTIAL INSTABILITY

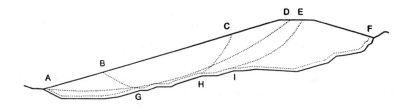

CATEGORY 1: NON-CIRCULAR SURFACE

     A-G-H-I-F

     [PASSING ENTIRELY ALONG LINER SYSTEM]

CATEGORY 2: NON-CIRCULAR SURFACES

     A-G-H-C, A-G-H-I-E, B-G-H-I-E

     [PASSING ALONG A PORTION OF LINER SYSTEM]

CATEGORY 3: CIRCULAR SURFACE

     A-D

     [PASSING ALMOST ENTIRELY THROUGH WASTE]

Figure 6.   Waste Mass Failure Modes

Response analyses for the 0.16 g, magnitude 7.7 MPE always showed amplification at the top of the landfill. Amplification factors as high as 3.5 were calculated at the top of the cover. Amplification of the average acceleration of the waste mass was also observed, although these amplification factors were less than the cover amplification factors, never exceeding 1.5. The greatest amplification for the cover usually, but not always, occurred for the thickest (90 m) waste column while the greatest amount of amplification for the waste mass always occurred for the thinnest (15 m) waste column. These results are consistent with the findings of Bray et al. (1995) for waste mass amplification and of Kavazanjian and Matasović (1995) for cover amplification.

The average acceleration time histories for the waste and cover system from the site response analyses were used in Newmark (1965) deformation analyses to estimate permanent seismic deformations for the liner and cover. The Newmark analyses were conducted using the computer program YSLIP_C (Yan, 1991). Permanent seismic deformations were evaluated as a function of yield acceleration to aid in the selection of materials and preparation of construction specifications. Results of the permanent deformation analyses are presented in Figure 7 for both the liner and cover systems. In this figure, the upper bound of the analyses performed for the eleven time histories selected to represent the prescriptive Subtitle D and the three different waste mass thicknesses are plotted with the results of the most critical analyses (the analyses resulting in the largest permanent seismic deformations) for three waste thicknesses and the four time histories selected to represent the MPE. The pattern of the calculated permanent seismic deformations with respect to waste mass thickness was essentially the same as the pattern of results with respect to peak acceleration behavior.

Results of the seismic performance analyses showed that while the larger intensity, smaller magnitude prescriptive Subtitle D design earthquake clearly governed the seismic performance of the waste mass, the large magnitude, smaller intensity MPE was equally important with respect to performance of the cover. If the site had been slightly closer to the San Andreas fault, or if the landfill foundation was susceptible to local amplification of seismic motions, the larger magnitude, smaller intensity MPE may have governed cover performance. The results of the analyses were used to establish minimum acceptable interface shear strength values for inclusion in construction specifications for selection of appropriate materials for construction of the landfill.

CONCLUSION

Seismic hazard analyses were performed for the Eagle Mountain Landfill in accordance with both the prescriptive approach provided in the Subtitle D regulations and the alternative site-specific approach required by California regulations. For the prescriptive Subtitle D approach, the MHA was evaluated from the most recent

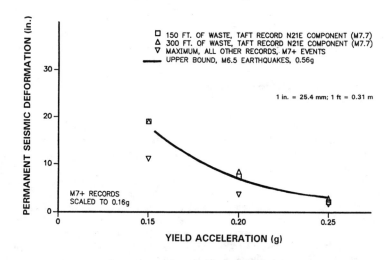

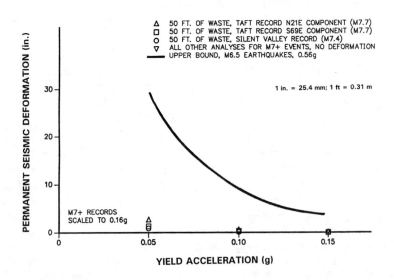

Figure 7.   Permanent Seismic Deformations of Waste Mass and Landfill Cover

version of the Algermissen maps published by USGS. A site-specific probabilistic analysis was performed in conjunction the prescriptive Subtitle D approach to validate the MHA interpolated from the Algermissen map and to establish an appropriate magnitude to associate with the MHA. Results of the analysis indicated that the design event evaluated in accordance with the prescriptive Subtitle D regulations was a magnitude 6.0 to 6.5 event approximately 8 km from the site generating a free field bedrock PGA at the site of 0.56 g. This event was not associated with any known fault but was instead attributed to the random seismicity assigned to the Southeast Transverse Ranges seismo-tectonic province.

In the alternative approach to evaluating the design earthquake required by California regulations, a site-specific analysis was performed to establish the MPE in accordance with CDMG guidelines. Results of this analysis indicated that the MPE was a magnitude 7.7 event on the Coachella Valley segment of the San Andreas fault at a distance of 53 km generating a free field PGA of 0.14 g at the site. The site-specific analysis also established a magnitude 8.0 MCE event due to simultaneous rupture of three segments of the San Andreas fault generating a free field bedrock PGA of 0.16 g at the site.

Suites of representative time histories were assigned to both the prescriptive Subtitle D and alternative California design events. For the MPE, the selected time histories were scaled to the MCE PGA due to a deficiency in the magnitude of some of the records and to provide a conservative basis for design. Landfill performance analyses were conducted using these records to establish minimum acceptable interface shear strength values for inclusion in construction specifications for selection of appropriate materials for construction of the landfill.

Evaluation of the results of the analyses described herein indicates that, while the prescriptive Subtitle D approach is different from the alternative approach employed in California, neither prescriptive nor alternative approach can be unconditionally described as more or less stringent than the other. In some cases, the prescriptive approach may result in larger design accelerations and calculated permanent displacements while in other cases the alternative approach may yield the maximum landfill response. Both approaches are considered to form the basis for design of a landfill with a high degree of seismic resistance that provides a high level of protection to the environment.

REFERENCES

Algermissen, S.T., Perkins, D.M., Thenhaus, P.C., Hanson, S.L. and Bender, B.L., (1990) "Probabilistic Earthquake Acceleration and Velocity Maps for the United States and Puerto Rico," United States Geological Survey Miscellaneous Field Studies Map MF-2120, Reston, Virginia.

Bender, B. and Perkins, D.M. (1987) "SEISRISK III, A Computer Program for Seismic Hazards Estimation," United States Geological Survey Bulletin 1772.

Bray, J.D., Augello, A.J., Leonards, G.A., Repetto, P.C., and Byrne, R.J. (1995) "Seismic Stability Procedures for Solid-Waste Landfills," *Journal of the Geotechnical Engineering Division*, ASCE, Vol. 121, No. 2. pp. 139-151.

CDMG (1975), "Recommended Guidelines for Determining the Maximum Credible and the Maximum Probable Earthquakes," *Technical Note No. 43*, California Division of Mines and Geology, Sacramento, California, 1 p.

Earth Mechanics (1992) "*Probabilistic Seismic Hazard Assessment - Eagle Mountain Landfill*," Technical Report, Earth Mechanics Inc., Fountain Valley, California.

Earth Technology Corporation (1988) "*In-Place Stability of Landfill Slopes, Puente Hills Landfill, Los Angeles, California*", Report No. 88-614-1, submitted to Los Angeles County Sanitation District, Whittier, CA, April.

Given, D.D. Jones, L.M. Hutton, L.K. and Hartzell S. (1987) "*The Southern California Network Bulletin, July through December, 1986*," United States Geological Survey Open File Report 87-488, 1987, 27p.

Jennings, C.W. (1992) "Preliminary Fault Activity Map of California," Division of Mines and Geology Open-File Report 92-03, Sacramento, California.

Joyner, W.B. and Boore, D.M. (1988) "Measurement, Characterization, and Prediction of Strong Ground Motion," *Proceedings of Earthquake Engineering and Soil Dynamics II - Recent Advances in Ground Motion Evaluation*," ASCE Special Geotechnical Publication No. 20.

Kavazanjian, E., Jr., and Matasović, N. (1995) "Seismic Analysis of Solid Waste Landfills," *Proceedings of Geoenvironment 2000*, ASCE Special Geotechnical Publication No. 46, pp. 1166-1180.

Martin, G.R. (1992) "Eagle Mountain Landfill Probabilistic Seismic Hazard Analysis," Technical Report prepared for GeoSyntec Consultants by Dr. Geoffrey R. Martin, Consulting Engineer, Palos Verdes, California

Newmark, N.M. (1965) "Effects of Earthquakes on Dams and Embankments", *Geotechnique*, Vol. 15, No. 2, June, pp. 139-159.

Proctor, R.J. (1992) "Faults and Micro-Seismicity Investigation and Conclusions, Proposed Eagle Mountain Landfill Site, Riverside County, California", Technical Report prepared for the Mine Reclamation Corporation by Richard J. Proctor, Engineering Geologist, Arcadia, California

Shlemon, R.J. (1992) "Geomorphic and Soil-Stratigraphic Age Assessments, Alluvial Deposits, Proposed Eagle Mountain Landfill Site, Riverside County, California", Technical Report prepared for the Mine Reclamation Corporation by Roy J. Shlemon and Associates, Newport Beach, California

Schnabel, P.B., Lysmer, J., and Seed, H.B. (1972) "*SHAKE: A Computer Program for Earthquake Response Analysis of Horizontally Layered Sites*", Report No. UCB/EERC-72/12, Earthquake Engineering Research Center, University of California, Berkeley, CA, December, 102 p.

Schell, B.A. (1992) "Seismotectonic Evaluation of the Proposed Eagle Mountain Landfill Region, Southern California," Report prepared by Bruce A. Schell, Consulting Geologist, Long Beach, California.

Sieh, K.E. (1978) "Pre-historic Large Earthquakes Produced by a Slip on the San Andreas Fault at Pallett Creek, California", *Journal of Geophysical Research*, Vol. 83, pp. 3907-3939.

Weldon, R.J., and K.E. Sieh (1985) "Holocene Rate of Slip and Tentative Recurrence Interval for Large Earthquakes on the San Andreas Fault in Cajon Pass, Southern California", *Geological Society of America Bulletin*, Vol. 96, pp. 793-812.

WGPFLESC (1992) "Future Seismic Hazards in Southern California, Phase I: Implications of the 1992 Landers Earthquake Sequence", Working Group on the Probabilities of Future Large Earthquakes in Southern California, *Southern California Earthquake Center Special Report*.

USEPA (1992) "Draft Technical Manual for Solid Waste Disposal Facility Criteria - 40 CFR Part 258," United States Environmental Protection Agency, April.

USEPA (1995) "RCRA Subtitle D (258) Seismic Design Guidance for Municipal Solid Waste Landfill Facilities," EPA/600/R-95/051, United States Environmental Protection Agency, April.

Yan, L.P. (1991) "*Seismic Deformation Analysis of Earth Dams: A Simplified Method*," Research Report No. SML 91-01, California Institute of Technology Soil Mechanics Laboratory, Pasadena, California.

Classification of Landfills for Seismic Stability Assessment

Ellis L. Krinitzsky, Mary Ellen Hynes,
Arley G. Franklin[1]

ABSTRACT

Earthquake ground motions at municipal solid waste landfills must be specified according to the level of hazard or criticality of the site along with the type of engineering analysis that is to be performed. The hazard is graded as (1) None to Negligible, (2) Low, (3) Moderate, and (4) Great. For non-critical sites, motions can be obtained from probabilistic maps as part of a non-site specific investigation: for critical sites, a deterministic, site specific evaluation is preferred and motions must be specified appropriately for the type of analysis, whether it is for foundation liquefaction, stability of slopes, integrity of barriers, earth pressures, or the design of appurtenant structures.

1. INTRODUCTION

During the past two decades, geotechnical earthquake engineers have evolved a consensus on a general approach to the evaluation of the stability of embankments under earthquake loading conditions. In broad terms, the approach follows the procedure shown in Figure 1. This approach has very general applicability, extending to sanitary landfills, and engineers working with seismic evaluations of landfills generally consider the specific analytical techniques used in the evaluation of earth structures to be applicable to sanitary landfills. These include conventional slope stability analyses, one-, two-, and three-dimensional dynamic response analyses, Newmark sliding block analyses, and others. However, the properties of the materials in sanitary landfills differ

---

[1]U.S. Army Engineer Waterways Experiment Station,
Vicksburg, Mississippi 39180

in significant ways from soils, and other characteristics of the "structure" (the landfill) differ in significant ways from conventional embankment or earthfill structures. These differences must be considered in evaluating the propagation of ground motions in the structure as well as the potential modes of failure. The observed performance of sanitary landfills in past earthquakes, considered in the light of observed and measured material properties and other characteristics of landfills, contributes significantly to our understanding of the potential modes of failure.

The seismic evaluation approach shown in Figure 1 is familiar to geotechnical earthquake engineers who deal with seismic effects on dams or other earthfill structures, and need not be elaborated upon in this paper. The paper will instead briefly discuss those characteristics that make the behavior of sanitary landfills differ from that of soil structures, and present the authors' recommendations for seismic hazard assessments at municipal solid waste landfills (MSWL) in accordance with U.S. Federal regulatory requirements for seismic safety (CFR 40, Parts 258.13 to 258.15). Included are methods to obtain earthquake ground motions and to assess earthquake effects on foundations, landfills, liners, cover, gathering systems for leachates and gases, and appurtenant structures.

2. CHARACTERISTICS OF SANITARY LANDFILLS

Composition and Structure

The nature of the waste materials in landfills varies widely in accordance with the source; i.e. industrial wastes of various kinds or the collected wastes of households and small businesses. The composition of household wastes is culture-dependent, but in more industrialized cultures it may be expected to have higher proportions of paper and packaging materials. From samples of 11 U.S. landfills, Rathje (1991) estimates the average composition as:

Type of waste	Average percent
Paper products	50
Organic; includes wood, yard waste, food scraps	13
Plastic	10
Metal	6
Glass	1
Miscellaneous; includes construction and demolition debris, tires, textiles, rubber, disposable diapers	20

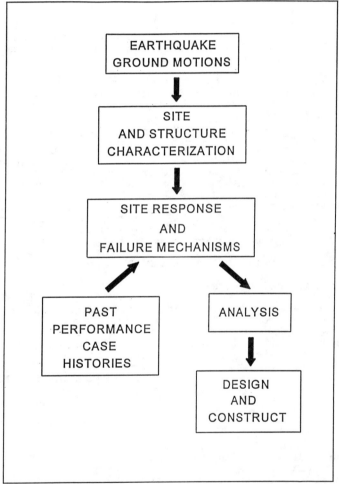

Figure 1. Seismic Stability Evaluation of Embankment or Earthfill Structures

In typical U.S. practice, the waste material placed during a working day is covered by a layer of soil at the end of the day, resulting in a cellular structure: discontinuous cells of waste materials separated by continuous cell walls of soil. There may also be membranes of geotextile or impervious plastic for the purposes of improving the strength of the fill or of controlling water movement. This cellular structure is of major significance in the stability of the landfill, since failure surfaces may tend to run through the soil layers or along the interfaces of soil or waste with synthetic membranes or between synthetic components.

The presence of toxic or otherwise hazardous materials in some landfills creates special performance requirements, i.e. preservation of the integrity of barriers to migration or release of those materials.

Decomposition

Because of the usually significant fraction of high-energy organic matter in sanitary landfills, significant biochemical activity is the rule. This results in chemical breakdown of both solid and liquid constituents to liquid and vapor end products. Much of this activity takes place under anaerobic conditions, and methane gas is a common product. Rates of decomposition are highly variable; Chen, et al., (1977) found reported rates of decomposition varying from 0.012 to 0.788 (by weight)/ year. However, much of the material, especially the paper products, may endure for surprisingly long times. In some excavated landfills, strata dating to 50 years or more before the present may contain readable newspapers and readily recognizable food items.

Compressibility and Settlement

According to Edil, et al. (1990), the primary mechanisms involved in the settlement in refuse fills are:

1. Mechanical (distortion, bending, crushing, and reorientation; similar to consolidation of organic soils);

2. Ravelling (movement of fines into large voids);

3. Physical-chemical change (corrosion, oxidation and combustion); and

4. Bio-chemical decomposition (fermentation and decay, both aerobic and anaerobic processes).

The authors observed that the rate of settlement was best described by the power creep law

$$S = H \, \Delta\sigma \, m \, t^n \qquad\qquad (1)$$

where

S = settlement
H = initial height of refuse
$\Delta\sigma$ = compressive stress, kPa
m = reference compressibility, 1/kPa
t = time, days
n = rate of compression.

Average values of the empirical parameters m and n obtained from observations of four sites were about 2.5×10^{-5} and 0.65, respectively. Values of m were highly variable, from 7.52×10^{-8} to 1.38×10^{-4}; values of n were less so, varying from 0.264 to 1.17. It has been observed by Edil et al. (1990), and by others, that primary consolidation, i.e. the dissipation of excess pore pressure, is largely complete at the end of placement of the refuse, and that creep dominates the compression behavior afterwards. Chen, et al. (1977) performed consolidation tests on laboratory specimens of saturated milled refuse materials, 2 to 4 inches (5 to 10 cm) in thickness, and found the pore pressure dissipation phase was generally completed in times of one minute or less, and that the settlement thereafter could be described by a creep model.

Density

Density of refuse material varies with composition, depth (overburden pressure), method of placement, and age. Refuse materials experience primary consolidation within the first few days of placement, but continue to settle and densify with time. Decomposition appears to have a negligible effect on density over a period of several decades. Measurements of density have been made from large-scale excavations in landfills, attempts at obtaining undisturbed samples from drill holes, and from laboratory consolidation tests used to simulate in situ materials and stresses. Landva and Clark (1990) developed a procedure for estimating the in situ unit weight of refuse based on the average unit weights of the constituents and the results of consolidation tests. Figure 2 shows recommended values of density for refuse as a function of depth from Kavazanjian et al. (1995) and Landva (1994). Although variation in density typically has little effect on the response of level ground sites and embankments, the dramatic variation in density for landfills has a distinct effect on dynamic response and is important to account for if a dynamic analysis is performed.

Shear Strength

The shear strength of landfill materials is difficult to measure since large-scale testing facilities are needed to accommodate a reasonable sample of the refuse and testing of actual refuse material may require special safety measures. The shear strength of refuse in landfills has been estimated by means such as back-calculation of stable landfill slopes (Kavazanjian et al. 1995), loading of fills in situ (Pagotto and Rimoldi 1987; Richardson and Reynolds 1991), and large-scale laboratory testing of refuse (Landva and Clark 1990, Landva 1994). Figure 3 shows a recommended shear strength for municipal solid waste materials based on these data: $c = 24$ kPa for normal stresses less than 25 kPa and $c = 5$ kPa with $\phi = 33$ degrees for normal stresses greater than 25 kPa. Landva (1994) recommends a lower range of $c = 0$ and $\phi = 33$ degrees based on direct shear tests. In the field, large landfill slopes are stable at slopes as steep as 1.2:1 to about 2:1, indicating strengths of at least $c = 5$ kPa and $\phi = 28$ degrees (Kavazanjian et al. 1995).

Dynamic Response

The dynamic response of a landfill depends on the nature of the incoming motions, landfill geometry, foundation conditions, stratigraphy or internal structure, density distribution, internal fluid pressures, moduli distribution, and moduli degradation with cyclic loading. This section focuses on in situ measurement of shear wave velocities for moduli estimation and back-calculation of modulus degradation from actual strong motion instrument recordings at a landfill.

Shear Wave Velocity. In situ measurement of shear wave velocities have been made using surface (spectral analysis of surface waves, seismic refraction) and/or subsurface techniques (down-hole, cross-hole, OYO in-hole) at several landfills as shown in Figure 4 (after Kavazanjian et al. 1995). These data indicate that shear wave velocities increase significantly with depth, by a factor of about two over a depth of 100 ft (30 m).

Modulus Degradation and Damping. Only one landfill in the United States has been instrumented with strong motion recording equipment, namely the Operating Industries, Inc. (OII) landfill located in Monterey Park, California. The OII landfill is a significant structure, with fill heights of greater than 250 ft (76 m), total

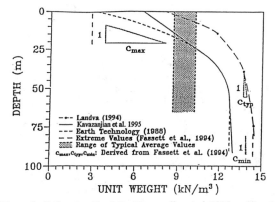

Figure 2. Refuse densities (after Kavazanjian et al. 1995, and Landva 1994)

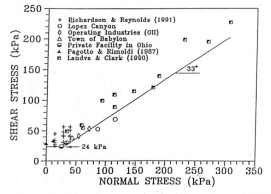

Figure 3. Refuse shear strength (after Kavazanjian et al. 1995)

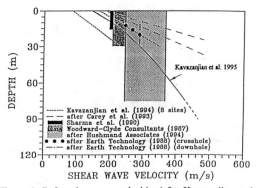

Figure 4. Refuse shear wave velocities (after Kavazanjian et al. 1995)

area of more than 1 square mile (2600 m^2) and slopes of about 2H:1V. Several recordings of earthquakes have been made with the instruments at OII. The 1994 Northridge, California Earthquake produced the largest motions yet recorded at the site. These recordings have been used to estimate modulus reduction and damping curves to use in equivalent linear dynamic response analyses. The results of back-calculations by Kavazanjian et al. (1995) indicate that the modulus reduction and damping curves for a refuse fill are similar to those for sand.

Performance of Landfills in Past Earthquakes

Anderson and Kavazanjian (1995) summarized observations from field inspections of landfills following three recent California earthquakes: the 1987 M 6.1 Whittier Narrows event, the 1989 M 7.1 Loma Prieta event, and the 1994 M 6.7 Northridge event. For the latter two earthquakes there is an abundance of data on landfill performance, since the earthquakes affected a wide area and numerous landfills were subjected to moderate to severe shaking.

In the Whittier Narrows earthquake, six landfills are known to have been subjected to estimated ground motions in the range of 0.10 to 0.30 g. The OII landfill was located only 4 km from the epicenter, and underwent an estimated free-field PGA of 0.30 g. Following the earthquake, some ground cracking in cover soils was reported at the OII landfill. No damage was reported from any of the others. At three of them, however, there were no post-earthquake inspections reported.

Following the 1989 Loma Prieta earthquake, there were reports of inspections of at least 17 landfills. One of these was 75 m high and subjected to an estimated PGA of 0.5 g. Others up to 45 m high experienced shaking with PGA's from 0.04 to 0.5 g. Only minor damage was reported, consisting typically of cracks 25 to 75 mm wide, occurring mostly at transition zones and at contacts between different materials, e.g. solid waste and natural ground. At one site, minor downslope movement of cover soils was noted.

Reports of investigations of 22 landfills are available following the 1994 Northridge earthquake. Of these, 16 were reported to have experienced motions exceeding 0.24 g (free-field PGA at base), and 6 to have experienced motions exceeding 0.38 g. The most severe damage reported was at the Chiquita Canyon landfill, 26 km from the epicenter, where the PGA was estimated to have been 0.39 g. Two tears were found in geomembrane liners, one 23 m long and one 3 m long. There was also cracking of cover soils, but no disruption of underlying impervious soil liners. In 17 cases the damage was categorized as minor or none. Where damage occurred, it typically consisted of superficial brittle cracking of cover soils and damage to gas recovery systems, including loss of power and breakage of gas and condensate lines and well heads. In all cases, the systems were back in operation within 24 hours.

3. SEISMIC HAZARDS

Seismically-Induced Permanent Displacement

As discussed above, the waste materials in sanitary landfills in general have relatively high shear strength, and experiences of landfills during past earthquakes have not produced any reports of large-scale movement or sliding in the landfill materials themselves. The most serious concern appears to be cracking of liner systems and covers and damage to appurtenances such as gas collection systems. Relatively small movements can pose a threat to the integrity of these systems. According to Anderson and Kavazanjian (1995) permanent displacements of 10 to 30 cm can be tolerated before these threats become serious.

Earthquake engineers generally accept the use of the Newmark sliding block analysis to estimate the potential permanent displacement in an embankment where there is no liquefaction threat. Description of the method, and computed displacements for over three hundred earthquake strong-motion time histories, are given by Franklin and Chang (1977) and Hynes-Griffin and Franklin (1984).

Hynes-Griffin and Franklin used a criterion of 1 m as an upper bound of permanent displacement in earth embankments to arrive at a recommendation for the use of a pseudostatic stability analysis with a seismic coefficient of one-half of the peak base acceleration, and a limiting factor of safety of unity, as a conservative screening method for permanent displacements. The authors have applied similar reasoning to sliding on deep surfaces in landfills, with the following modifications: (1) the mean $+\sigma$ curve for displacements (Hynes-Griffin and Franklin 1984) is used, rather than the upper bound for all records analyzed; (2) the amplification factor for ground motions propagated into the embankment is graded from upper bound to mean values (Hynes-Griffin and Franklin 1984) as the base motions increase; and (3) a displacement limit of 1 foot (30 cm) is used. This analysis indicates that a pseudostatic analysis with a seismic coefficient of 0.2 × PGA + 0.1 g, with a factor of safety of unity, is suitable for screening against permanent displacement for a deep-seated failure surface. Displacement analyses using a Makdisi and Seed (1978) approach also support a screening seismic coefficient of 0.2 × PGA + 0.1 g for deep-seated surfaces.

This analysis also indicates that if the static factor of safety against slope instability is at least 1.5, then this mode of failure is not of concern for ground motions less than 0.15 to 0.20 g. But considering other mechanisms of seismically induced deformation or settlement leads us to believe a range of 0.10 to 0.15 g is prudent as a threshold below which analysis is not needed.

For permanent displacements near the crest of a landfill involving cover materials, simplified Newmark sliding block analyses either by the Hynes-Griffin and Franklin (1984) approach or by the Makdisi and Seed (1978) approach indicate damaging levels of displacement may occur at low levels of ground motion, ≤ 0.1 g. We believe that cover materials should either by designed so that they can easily be repaired or special features will need to be designed into the cover systems to increase yield acceleration values.

Liquefaction in Foundation Soils

The fibrous materials that dominate in municipal landfills are not themselves susceptible to liquefaction. However, a susceptibility to liquefaction in foundation soils can have a dangerous potential for inducing differential displacement in a landfill. Evaluation of this hazard is discussed in detail by Koester and Franklin (1985) and Carter and Seed (1988).

Fault-Induced Permanent Displacement

MSWL units should not, and under U.S. federal regulations may not, be located within 200 feet (60 meters) of an active fault, i.e. one evidencing Holocene displacement, unless the fault can be proven not to present a threat. The concern is that a fault movement beneath a landfill can result in a permanent displacement for which the fill was not designed. An active fault is considered to be one that is likely to move again with or without an earthquake. Movement will be either along an identifiable trace or within a zone in which fault movement has occurred in the past.

Existing landfills may not have been evaluated for fault hazards. Reexamination of a site should include

potentials for fault movement and the extent to which
damages may occur.

Karst Instability

In areas where the bedrock is limestone or other soluble
rock, sudden collapse of the ground into solution
cavities can pose a special hazard. Such collapses are
known to have taken place as a result of earthquake
shaking. Not only would they be a cause of displacements
in fills, but they are invariably connected with
groundwater aquifers and pose a threat of contamination.

Activation from Mining, Fluid Extraction, etc.

Similar to the problems with karst, the presence of man-
made cavities from mining or other activities may
destabilize the foundation of a landfill. Extractions of
fluids such as groundwater or petroleum also may cause
differential settlements. Fluid extraction has an effect
of causing shallow non-seismic activation of movement
along existing fault planes.

*Susceptibility to Landslides, Avalanches, and Debris
Flows*

Landfills often are situated in rugged terrain that is
unsuited to other uses. Such areas may be prone to having
landslides, avalanches, land slips, and mud flows or
debris flows.

The above hazards need to be assessed as part of the
general geological evaluation to assure stability at a
landfill site. In some cases they can be done routinely,
simply by noting their absence as threats to a site. For
others, there are no generally accepted procedures. At
potentially hazardous sites the investigation should be
done in a way that is defensible for each set of
circumstances.

4. CLASSIFICATION AND RECOMMENDATIONS FOR ANALYSIS

Seismicity

Federal regulatory guidance (CFR 40, Part 258.14)
provides for two methods of investigation of the
potential ground motions at the site: the first, which
may be termed a non-site-specific evaluation, uses a
seismic hazard map (Algermissen, et al. 1990) showing
mean values of acceleration on rock with a 90 percent
probability of non-exceedance in 250 years.

The 250-year map cannot be site-specific because it must generalize information that is both unevenly distributed and of greatly uneven quality. Microzoning for sensitive local conditions is mostly neglected. Motions assigned very often are _effective_ motions, those that eliminate peak values judged to have the smallest likelihood of being experienced. Effective motions attempt, sometimes crudely, to reduce costs in construction for earthquakes that are uncertain. A critical landfill in an area of seismic threat needs a site-specific evaluation that is tightly focused on conditions pertinent to the local environs.

The second method permits an investigator to perform an analysis by which is obtained the maximum expected horizontal acceleration based on a site-specific seismic evaluation. A site-specific evaluation requires a geological-seismological study that starts at the beginning, reviews all relevant data, defines the earthquake source areas that are meaningful for the site, assigns magnitudes, attenuations, etc., and makes an independent designation of site-specific ground motions. In general site-specific evaluations may be done by either deterministic or probabilistic methods. At a critical site, the design should accommodate the worst that can reasonably be expected regardless of when it may happen; probability should not be a determining consideration.

A site-specific evaluation may indicate several earthquake sources that can affect a site. Each significant source should be analyzed with its own suite of accelerograms.

In both cases, a Design Level Earthquake (DLE), or earthquakes, are assigned. The DLE may be a mean value or a mean +σ value depending on the severity of the hazard and the type of engineering analysis, as indicated in Tables 1 and 2.

Levels of criticality

The level of criticality of a site determines the minimum extent of investigation that the site requires, whereas economic considerations normally tend to drive us toward minimizing the investigative effort. We thus focus on the minimum levels of investigation required and the minimum threshold values of ground motion that would trigger the requirement for various kinds of investigations. However, the level of criticality is often a subjective judgment. We approach the issue by asking the question, what are the consequences of failure? If they are intolerable, then the site is a

Table 1. Relation of Landfill Problem to
Grade of Seismic Hazard

Landfill Problem	Grade of Seismic Hazard*	
	Non-Critical**	Critical***
A. Geometry susceptible to deformation	1,2,3	3,4
B. Presence of toxic substances	1,2	3,4
C. Danger to the environment, ground water, other	1,2	3,4
D. Danger to population	1,2	3,4
E. Danger to lifelines, transportation, other	1,2,3	4

* 1 = None to negligible
  2 = Low
  3 = Moderate
  4 = Great

** Consequences of failure are tolerable. Damage readily repaired.

*** Consequences of failure are intolerable.

Table 2. Grade of Seismic Hazard with Threshold Acceleration and Corresponding Engineering Analysis

Grade of Seismic Hazard	Seismic Impact Zone CFR 40, Part 258.14	Initial Earthquake Hazard Evaluation	Type of Hazard	Threshold for PHA on Rock and Corresponding Analysis Method
1 (None to Negligible)		Non-site specific: 250-yr probability map*, 90 percent probability of nonexceedance. Check against deterministic information for worst-case scenario.	<0.10 g and No liquefaction	No analysis.
2 (Low)	≥0.10 g Mean horizontal motion on rock. 250 year, 90 percent probability of nonexceedance.	Non-site specific: 250-yr probability map. Compare deterministic worst case. Perform geological reconnaissance. Site specific: Only if site is susceptible to liquefaction.		
3 (Moderate)		Site specific: Deterministic investigation: Locate earthquake sources, assign maximum credible earthquakes for each source, attenuate motions to the site, specify free field motions for the site, select analogous time histories for excitation at the site.	a. Foundation materials susceptible to liquefaction.	a. For liquefaction: ≥0.05 g to 0.10 g (mean + σ); Perform dynamic analysis.
4 (Great)			b. Possible damage to liner, cover, or gathering system for leachates or gases. Possible deformation within the landfill gathering system for leachates or gases. Possible deformation within landfill.	b. For sliding and deformation: (1) ≥0.10 g to 0.15 g (mean + σ); Perform pseudostatic analysis. (2) ≥0.15 g (mean + σ); Perform dynamic analysis.

* Algermissen, et al. 1990.

critical one. If they are tolerable, then the site is not
critical.   At non-critical sites, the consequences of
failure will be essentially economic losses, which could
range from inconsequential to heavy.

Grade of Seismic Hazard

The   U.S. federal   regulatory   guidance   (CFR 40,
Part 258.15) requires that potential instabilities in the
foundation of the landfill, or that of adjacent areas
from which the landfill might be affected, be evaluated.
The landfill must be shown to present no hazard from
instability at the site, or is adequately protected
against whatever instability may exist.   The hazards can
include unstable foundation soils, nearby landslides or
other mass movement, weakness from underlying karst,
differential displacement from fluid extraction, collapse
in mines or from other underlying cavities, or whatever
else may be a source of potential instability at the site.

The guidance (CFR 40, Part 258.14) establishes a basic
threshold at which an investigation is mandated: this
corresponds to a 90 percent probability that a ground
motion of 0.10 g will not be exceeded in 250 years.   The
extent of the investigation that is appropriate needs to
be determined by the grade level of the seismic threat.

The potential hazards to the environment and to people
due to the presence of MSWL can be highly variable.   The
possible effects depend on the geometry of the fill and
its susceptibility to deformation; the presence of toxic
substances, with problems of fluid leakage and emission
of gases; danger to the environment due to contamination
of groundwater, surface waters, etc; direct threats to
populations, and danger to lifelines, transportation, and
other essential facilities.

Four grades of seismic hazard are used: (1) None to
negligible,   (2) Low,   (3) Moderate,   and   (4) Great.
Assigning a grade to the seismic hazard is essentially
subjective, but it involves consideration of the site
seismicity,   the   criticality   of   the   site,   and   the
potential for loss or injury, including economic losses
at non-critical sites.   Grading these factors can be
either straightforward or highly subjective.   Grading
needs to be done on a site-specific basis.   There are no
rules.

Table 1 presents ranges of the grade of seismic hazard
associated with the type of threat, or problem, and the
criticality of the site.   Table 2 gives an approximate
relationship between the nature of the hazard and the
grade levels of Table 1, and acceleration thresholds and

corresponding seismic evaluations that are appropriate for each grade.

The levels of grade discrimination in Table 2 are categorized by whether a site is critical or non-critical. A non-critical site may be evaluated using seismic probability maps in a non-site specific appraisal, provided there is no susceptibility to liquefaction. A critical site requires a site specific evaluation and one should be guided by thresholds of earthquake ground motion that are appropriate for the kind of engineering analysis being considered. A site that is susceptible to liquefaction, though it is otherwise non-critical, may in some cases require a site specific evaluation. The general rule is that for critical sites the requirements for analysis and design are governed by considerations of public safety; for non-critical sites economic considerations generally govern, and one must balance the present costs of investigations, analysis and earthquake-resistant design against the potential future (and therefore discounted) losses due to earthquake damage.

Pseudostatic Analyses

Earthquake ground motions appropriate for various categories of use in pseudostatic analyses are shown in Table 3. These are based on criticality of the landfill and the seismicity level of the area. The analyses are for stability of slopes, permanent deformations and integrity of soil or membrane barriers, earth pressures on retaining walls, and structures of concrete and/or steel frame. The indicated motions, unless otherwise designated, are intended for use as seismic coefficients. As indicated in Table 3, pseudostatic analyses are not suitable for analysis of foundation liquefaction and are not suitable for motions greater than 0.40 g. Similarly, they are not suitable for analysis of slopes, soil barriers, or earth pressures where the soils involved may be susceptible to liquefaction.

Dynamic Analyses

Earthquake ground motions for use in dynamic analyses are shown in Table 4. This type of analysis is indicated for a critical landfill in an area of moderate to strong seismicity. A threshold acceleration of 0.15 g (mean + σ) on rock is appropriate to analyze deep-seated sliding or deformation. Susceptibility to liquefaction, in any area, requires a threshold for motions as low as 0.05 g on rock. These motions serve as parameters for shaping time histories to provide cyclic shaking for use in the

Table 3. Earthquake Ground Motions for Use in Pseudostatic Analyses

	Foundation Liquefaction	Stability of Slopes and Integrity of Barriers	Earth Pressures	Concrete and/or Steel Frame Structures
• Non-critical landfill in any zone of seismic activity. • Critical landfill in an area of low seismicity (Mean + σ PHA <0.15 g) • Obtain <u>Design Level Earthquake(s)</u> (DLEs)	Pseudostatic analyses do not apply. Use dynamic analyses.	• Use $0.2 \times$ PHA $+ 0.1$ g as seismic coefficient, F.S.≥1.0. • PHA is obtained from (mean)* values: (a) MM intensity (b) Magnitude-distance attenuation (c) 250-yr, 90 percent probability of nonexceedance.	• PHA from (mean)* values: (a) MM intensity (b) Magnitude-distance attenuation (c) 250-yr, 90 percent probability of nonexceedance. • Use 1/2 PHA for backfill.	• Seismic-zone coefficients/factors in building codes. • For generating ratio of PHA to A of structure or element, PHA is obtained from (mean)* values: (a) MM intensity (b) Magnitude-distance attenuation (c) 250-yr, 90 percent probability of nonexceedance.
• Critical landfill in an area of moderate to strong seismicity (Mean + σ PHA ≥0.15 g ≤0.40 g) • Obtain <u>Design Level Earthquake(s)</u> (DLEs)	Use dynamic analyses.	• Use $0.2 \times$ PHA $+ 0.1$ g as seismic coefficient, F.S.≥1.0. • PHA from (mean + σ)* values: (a) MM intensity (b) Magnitude-distance attenuation (c) 250-yr, 90 percent probability of nonexceedance.	• PHA from (mean + σ)* values: (a) MM intensity (b) Magnitude-distance attenuation (c) 250-yr, 90 percent probability of nonexceedance. • Use 3/4 PHA for backfill.	• Seismic-zone coefficients/factors in building codes. • PHA from (mean + σ)* values: (a) MM intensity (b) Magnitude-distance attenuation (c) 250-yr, 90 percent probability of nonexceedance.

* Adjust if necessary for site condition; shallow plate boundary, deep subduction zone, or intraplate area; near field or far field; effective motions when near an earthquake source.

Note: PHA is the free-field Peak Horizontal Acceleration.

Table 4. Earthquake Ground Motions for Use in Dynamic Analyses

	Foundation Liquefaction	Stability of Slopes and Integrity of Barriers	Earth Pressures	Concrete and/or Steel Frame Structures
• Critical landfill in an area of moderate to strong seismicity for sliding and deformation: PHA ≥0.15 g (mean + σ) for liquefaction in any area: PHA >0.05 g (mean + σ) • Obtain <u>Design Level Earthquake(s)</u> (DLEs)	• Peak horizontal motions • Generate time histories	• Peak horizontal motions (mean + σ)* • Generate time histories.	• Peak horizontal motions (mean + σ)* • Generate time histories.	• Peak horizontal motions (mean + σ)* • Generate time histories. • Obtain response spectra for above time histories. • Alternatively, go directly to response spectra, entering with the above peak motions. • Check response at the natural frequency of the structure.

* Obtain peak horizontal motions from (a) MM intensity or (b) magnitude-distance attenuation charts. Adjust for site condition; shallow plate boundary, deep subduction zone, or intraplate area; near field or far field effective motions when near an earthquake source.

dynamic analyses.   The parameters are also levels for entrance into response spectra.

5. CONCLUSIONS

The level of sensitivity to seismic hazard, based subjectively on dangers to people and the environment, is graded as (1) None to Negligible, (2) Low, (3) Moderate, and (4) Great.   These are based on considerations of seismicity, the potential consequences of earthquake damage, and whether the site is considered to be Critical or Non-critical.   A non-critical site may appropriately be evaluated with a non-site specific study based on ground motions obtained from probabilistic maps and may require an engineering analysis only if there is a liquefaction threat or if the potential economic losses justify it.   A critical site in an earthquake-prone area requires site specific earthquake ground motions.   Levels of ground motion suitable for various types of engineering analysis, and thresholds calling for their use, are recommended for analysis of foundation liquefaction, stability of slopes, permanent displacements and integrity of barriers, earth pressures, and appurtenant structures.

6. ACKNOWLEDGEMENTS

The studies described, unless otherwise noted, were conducted under research programs of the United States Army Corps of Engineers.   Permission was granted by the Chief of Engineers to publish this information.   Opinions and recommendations presented in this paper are those of the authors, and do not represent official positions of the Corps of Engineers.

7. REFERENCES

Algermissen, S. T., Perkins, D. M., Thenhaus, P. C., Hanson, S. L., and Bender, B. L.   1990.   Probabilistic Earthquake Acceleration and Velocity Maps for the United States and Puerto Rico, U.S. Geological Survey Miscellaneous Field Studies Map MF-2120, 2 sheets, Washington, D.C.

Anderson, D. G. and Kavazanjian, E., Jr.   1995. "Performance of Landfills Under Seismic Loading," Proceedings, Third International Conference on Recent Advances in Geotechnical Earthquake Engineering and Soil Dynamics, Rolla, MO.

Carter, David P. and Seed, H. Bolton.   1988. "Liquefaction Potential of Sand Deposit Under Low Levels of Excitation," Report No. UCB/EERC-88/11, University of California at Berkeley.

Chen, W. H., Zimmerman, R. E., and Franklin, A. G. 1977. "Time Settlement Characteristics of Milled Urban Refuse," Proceedings, ASCE Conference on Geotechnical Practice for Disposal of Solid Waste Materials, Ann Arbor, MI, pp. 136-152.

Code of Federal Regulations (CFR) 40, Parts 258.13 to 258.15, Revised as of July 1, 1993. Federal Register, Washington, D.C.

Edil, T. B., Ranguette, V. J., and Wuellner, W. W. 1990. "Settlement of Municipal Refuse," Geotechnics of Waste Fills - Theory and Practice, ASTM STP 1070, Philadelphia, PA, pp. 225-239.

Franklin, A. G. and Chang, F. K. 1977. "Earthquake Resistance of Earth and Rock-Fill Dams; Permanent Displacements of Earth Embankments by Newmark Sliding Block Analysis," Miscellaneous Paper S-71-17, Report 5, U.S. Army Engineer Waterways Experiment Station, Vicksburg, MS.

Harder, L. F., Jr. 1991. "Performance of Earth Dams During the Loma Prieta Earthquake," Proceedings, Second International Conference on Recent Advances in Geotechnical Earthquake Engineering and Soil Dynamics, Vol. II, Shamsher Prakash, ed., Rolla, MO, pp. 1613-1629.

Hynes-Griffin, Mary E. and Franklin, Arley G. 1984. "Rationalizing the Seismic Coefficient Method," Miscellaneous Paper GL-84-13, U.S. Army Engineer Waterways Experiment Station, Vicksburg, MS.

Kavazanjian, Edward, Jr., and Matasovic, Neven. 1995. "Seismic Analysis of Solid Waste Landfills," Geoenvironment 2000, v. 2, ASCE Geotechnical Special Publication No. 46, pp 1066-1080.

Kavazanjian, Edward, Jr.; Matasovic, Neven; Bonaparte, Rudolph; Schmertmann, Gary R. 1995. "Evaluation of MSW Properties for Seismic Analysis," Geoenvironment 2000, v. 2, ASCE Geotechnical Special Publication No. 46, pp. 1126-1141.

Koester, J. P. and Franklin, A. G. 1985. "Current Methodologies for Assessing the Potential for Earthquake-Induced Liquefaction in Soils," Report No. NUREG/CR-4430, U.S. Nuclear Regulatory Commission, Washington, D.C.

Landva, A. O. 1994. Personal Communication.

Landva, A. O. and Clark, J. I. 1990. "Geotechnics of Waste Fill," <u>Geotechnics of Waste Fill - Theory and Practice</u>, STP 1070, ASTM, Philadelphia, PA.

Makdisi, F. I., and Seed, H. B. 1978. "Simplified Procedure for Estimating Dam and Embankment Earthquake Induced Deformations," <u>Journal of the Geotechnical Engineering Division</u>, ASCE, Vol. 104, No. GT7, pp. 849-867.

Pagotto, A., and Rimoldi, P. 1987. "Design and Construction of a Geogrid Reinforced Embankment Over Waste Material," <u>Proceedings of Geosynthetics '87</u>, New Orleans, LA.

Rathje, William L. 1991. "Once and Future Landfills," <u>National Geographic</u>, V. 179, No. 5, May 1991, pp. 116-134.

Richardson, G., and Reynolds, D. 1991. "Geosynthetic Considerations in a Landfill on Compressible Clays," <u>Proceedings of Geosynthetics '91</u>, Vol. 2, Atlanta, GA.

GROUND-MOTION CHARACTERIZATION FOR PROBABILISTIC RISK BASED SEISMIC EVALUATION AND DESIGN OF SOLID WASTE LANDFILLS

Robert T. Sewell, [1] Member, ASCE

Abstract

This paper provides the reader with guidance and insights on developing ground-motion input for performing a seismic design or a seismic evaluation of a solid waste landfill. The principal goal of this guidance is to enable the engineer or decision maker to achieve specific probabilistic performance objectives, in order to limit the risk of landfill failure and the risk of consequential environmental contamination. The guidance demonstrates the roles of both deterministic methods and probabilistic methods of analysis in the systematic development of risk-consistent motion input. The significance of this guidance for ground-motion input determination, with respect to safety decision making and performance-based regulatory criteria, is highlighted.

Introduction

Seismic Failure Concerns for Solid Waste Landfills. As for other types of earthen structures and embankments (e.g., dams, levees, fills, slopes), a solid waste landfill (SWL) may undergo significant deformations, when subjected to earthquake ground motion, potentially resulting in notable damage or failure. Such damage or failure for a SWL has the potential to lead to environmental contamination, and may result in adverse effects to the health of the nearby population and wildlife.

Significant components of a SWL that may be damaged in an earthquake include: the liner/barrier system (primary and secondary liners), the leachate collection system, the ground-water collection system, monitoring systems, the methane gas collection system (if present), and the permanent cover system (for a post-closure facility). In addition, earthquake damage may be isolated to the waste pile itself -- for example, due to waste-slope failure; such damage may be largely inconsequential, provided other systems are not effected and the ability of the SWL to perform its intended function is not impaired.

---

[1] President, R.T. Sewell Associates, Inc., 705 Delphi Drive, Lafayette, CO 80026

Failures of the various SWL components may occur due to excessive displacements along a developed failure/slip surface or due to excessive soil strains. Clearly, the most significant failure would involve a breach of the liner system, which may be difficult or impossible to repair satisfactorily with existing technology, and may well result in environmental contamination. Cracking of liner soil, relative displacement of the leachate collection layer, rupture of collector piping, etc., are some additional damage-related concerns associated with ground-motion loading.

Unique Considerations for Landfills. A unique characteristic of a SWL is the potential inherent weakness which may be introduced in the liner system (i.e., due to low-strength interface problems). Typically, the liner system is a composite layer constructed of geosynthetic (geomembrane, geonet, other geotextile) and compacted clay. This configuration can encourage the formation of a slip surface along the liner, or create a tendency for excessive cracking in the clay layer (which would compromise the performance of the liner), under earthquake loadings. Another characteristic substantially unique to SWLs is the continually changing geometry over the facility's active life (i.e., pre-closure period). Because the height and extent of the waste pile increase over time, there are a variety of possible dynamic-response configurations that are implied; these possible configurations need to be explicitly addressed in the design or evaluation process. And, of course, SWLs are also unique with respect to the variation and random character of fill materials and the resulting non-uniformity and uncertainty of fill properties (including shear modulus, unit weight, damping, etc.).

These unique characteristics for SWLs create challenges for satisfactory performance evaluation and performance-based design (e.g., see Martin and Kavazanjian, 1994; Mitchell and Mitchell, 1994; Repetto et al., 1994; and Seed and Bonaparte, 1994). A meaningful performance assessment must adequately model and address these unique elements of SWL behavior, using detailed analyses, as appropriate. Hence, even if the ground-motion input for performance evaluation of a SWL were determined in a simplified manner, performance assessment would still be a highly complex problem.

Ground Motion Considerations. Levels of maximum transitory deformations, of permanent seismic-induced displacements, or of other meaningful nonlinear-response measures that govern damage or failure of a SWL are highly sensitive to specific details of the input ground motion (e.g., motion non-stationarity and strong-motion duration). Therefore, the manner in which ground motion input is specified for SWL evaluation or design must be capable of meaningfully conveying the damage significance of expected strong motions. Furthermore, the motion should be characterized in a form readily amenable to engineering analysis. Hence, careful consideration clearly must be given to the ground-motion selection and characterization process, in order to achieve a meaningful seismic performance assessment for a SWL.

Discussion Outline. The remainder of this paper focuses on the specific aspect of systematic motion selection and characterization procedures for use in seismic performance assessment of SWLs. First, a background on ground-motion damage potential is provided; second, seismic performance goals for SWLs are discussed; third, an overview of risk-assessment considerations for performance-based seismic evaluation and design is presented; fourth, a proposed approach for ground-motion selection and characterization, for SWL performance-based evaluation or design, is described; fifth, the important roles of both deterministic and probabilistic methods in seismic performance assessment are clarified; sixth, facilitation of the SWL regulatory process (including responses to regulatory criteria by SWL owners and/or operators), through the proposed probabilistic risk-based performance assessment techniques, is highlighted; and lastly, a summary discussion and conclusions are provided with respect to the significance of probabilistic risk-based determination of ground-motion input for seismic evaluation or design studies of SWLs.

Characterizations of Motion Damage Potential

Motion Time Histories. The best engineering characterization of a given motion's potential to damage any type of structure, including a solid waste landfill, is its time history of ground acceleration. With this complete characterization, an engineer can conduct seismic response analyses of varying complexity, ranging from simple Newmark sliding block analyses to three-dimensional non-linear dynamic finite-element analysis. Hence, estimates of maximum displacements, permanent deformations, soil strains, etc., can be obtained and can account for the specific details (timing, magnitude, and number of significant cycles) of the motion input that are important to damage. Detailed time-history response analysis of a SWL, however, tends to be expensive, and thus, would generally be implemented sparingly.

Inelastic Response Spectra for SWLs. An inelastic response spectrum is the next best characterization of a given motion's damage potential. Inelastic spectra inherently convey a quantitative, engineering measure of the damage significance of important details (including non-stationarity and duration) of the motion time history. To be applicable to the seismic assessment of a SWL, however, inelastic response spectra must be constructed with the following special considerations in mind: (1) dynamic models that reasonably represent nonlinear response characteristics of SWLs need to be employed; and (2) measures of damage (e.g., maximum deformation, permanent deformation, etc.) that are meaningful to SWL performance need to be computed as the basis for inelastic spectra. Thus, as opposed to being based on the maximum ductility response of a single-degree-of-freedom oscillator with bilinear force-deformation behavior (as is typically the case for inelastic spectra developed for framed structures), the inelastic spectra for a SWL would be constructed, for example, based on the maximum permanent displacement response of a single-degree-of-freedom oscillator attached to a sliding block. Several time history response analyses (albeit simplified analyses) need to be conducted in order to define inelastic spectra for a single motion. However, once the inelastic spectra are obtained, they may be readily applied for assessment of a variety of SWL configurations.

Linear Response Spectrum. The linear response spectrum is the minimal characterization of motion damage potential that still has meaning for SWL performance; it directly conveys measures of a given motion's ability to induce linear dynamic response amplification in structures, as a function of predominant frequency. Furthermore, although a linear response spectrum is not directly based on analyses of nonlinear response and assessment of damage measures, it does, nonetheless, also indirectly convey important information on the ability of a motion to induce nonlinear damage. Specifically, inelastic spectral reduction factors (that are used to construct an inelastic response spectrum from a linear response spectrum) can be reasonably estimated (with a coefficient of variation of about 0.15 to 0.25) based solely on ordinates of linear response spectra (Kennedy et al., 1984). For conceptual simplification, it may be said that information pertaining to nonlinear damageability is conveyed by means of the slope of the linear response spectrum. Hence, if $(\Delta S_a/\Delta T)$ -- where S_a denotes linear spectral acceleration and T denotes period of vibration -- is a significant positive value at a period corresponding to the initial predominant response frequency of a SWL, then nonlinear damage potential (for a given linear response spectral ordinate) will be high; alternatively, if the slope $(\Delta S_a/\Delta T)$ is a significant negative value, then nonlinear damage potential will be low. This conceptual relation is based on the widely known understanding that, as any structure losses stiffness, its period of response lengthens, and hence, if a motion has increased strength over the range of such longer response periods, then increased dynamic response amplification will occur.

Peak Ground Acceleration. Peak ground acceleration (PGA) is useful as a characterization of motion damage potential only for very high-frequency structures. By itself, PGA is a poor ground-motion measure for engineering evaluation of lower-frequency structures, including SWLs. PGA has little correlation with the strength of motion input over the (predominant-response) frequency range of interest for SWLs. Thus, if one could choose only a single parameter to describe the strength of a given motion, in order to evaluate the motion's expected impact on a SWL, PGA would be an extremely poor choice. Inelastic spectral acceleration at the SWL's predominant response frequency, linear spectral acceleration averaged over frequencies varying from the initial frequency to secant frequency of the SWL, or even linear spectral acceleration at the SWL's predominant frequency would be much better choices to describe the damage potential of the given motion. Thus, PGA is not well suited as sole basis for determining seismic demands for use in SWL assessment, nor is it a very good decision parameter for developing siting criteria, characterizing seismic hazard, or establishing key seismic design guidelines.

The deficiencies of PGA are, however, substantially mitigated by use of an appropriate spectral shape for design or evaluation. In this case, PGA is simply used as a convenient scaling parameter, and the selection of a proper spectral shape still remains as the paramount consideration for seismic assessment purposes. Although this process (of scaling an appropriately determined spectral shape based on PGA) places undue (and misleading) emphasis on PGA as a motion parameter, it is acceptable only because the appropriate spectral demands are ultimately characterized. The reader should recognize, however, that the value of PGA itself, for any given real earthquake motion, has little relevance to the demands imposed on a SWL by that motion.

Probabilistic Considerations: Representations of Seismic Hazard. For probabilistic risk-based performance assessment, one must be able to associate probabilities with the levels of interest of the chosen characterization of motion damage potential. Such evaluation of probabilities, for any of the preceding motion characterizations, can be accomplished using probabilistic seismic hazard methodology, as will be discussed more fully later. At the most fundamental level, therefore, probabilities may be determined for magnitude-/distance-dependent motions, when magnitude-/distance-dependent time-history representations of the seismic loading is required. Where a spectral representation of the seismic loading is required, probabilistic linear response spectra may be developed, as presented in the form of uniform hazard spectra (UHS); similarly, probabilistic inelastic response spectra may be developed and presented in the form of constant-damage, uniform hazard spectra (CDUHS) [Sewell and Cornell, 1988]. And lastly, for cases where a site-specific spectral shape is specified without the benefit of probabilistic hazard analysis (e.g., as a median, average, or 84th-percentile spectral shape obtained from a suite of motions), and PGA is used as the spectral scaling parameter, a seismic hazard curve for PGA could be developed to approximately describe probabilities of exceedance of the implied motion spectra.

Seismic Performance Objectives

Deterministic and Probabilistic Performance Objectives. Deterministic performance objectives for seismic response of a SWL would logically aim at limiting seismic deformations, permanent displacements, soil strains, etc., to help insure that damage is acceptably low (e.g., does not occur) for a specified design or evaluation input motion. However, without special attention to the development of the input motion -- for example, if the input motion is evaluated solely on a deterministic basis -- nothing can be said concerning the likelihood that a given design will fail to achieve its safety function. More meaningful performance objectives, particularly for purposes of safety management and establishing regulatory criteria, would convey limits on probabilities of failure and probabilities of unacceptable levels of environmental contamination. Such probabilistic risk-based performance objectives help insure a substantial measure of consistency and equitableness in the safety management process. Without such risk-based performance goals, safety levels are vague; i.e., the public is not assured a safety standard that limits risk to an established level, and the SWL owner/operator is not assured of safety-consistent design requirements.

Use of Seismic Safety Goals for SWL Performance Assessment. Seismic performance objectives are thus most usefully stated as probabilistic safety goals. Specific safety goals for seismic performance of SWLs have not yet been established. Perhaps the most logical goal would define limits on the annual probability of seismic-induced breach of the liner system. In addition, assuming that a liner breach could not be detected or repaired prior to measurable release of harmful contaminants, it would be prudent to also define a limit on the frequency (and timing) of unacceptable ground-water contamination at all applicable control locations (e.g., drinking or irrigation water intake locations, nearby wetlands, etc.). For the sake of discussion and subsequent illustration, the following goals for design and evaluation of SWLs are proposed in this paper: (1) mean annual probability of seismic-induced liner breach not to exceed 5×10^{-4} per year; and (2) mean annual probability of unacceptable ground-water contamination at any control location not to exceed 5×10^{-5} per year. (As shown later, the first of these goals appears to be roughly consistent with current regulatory criteria for SWLs). The second of these goals requires ground-water transport and hydrogeological analyses to quantify, and thus, it is somewhat more complicated to assess than the first goal.

Use of risk-based performance goals has seen application, or proposed application, for assessment of a variety of engineering facilties (e.g., see EPRI, 1994; Kennedy et al., 1992; USNRC, 1986).

Use of Safety Goals for Input-Motion Determination. The subsequent approach to risk-based ground-motion determination presumes that relevant probabilistic safety goals have been established to define performance objectives. The approach does not depend on the specific value or criteria for the safety goal; setting such criteria is a matter of public policy and regulatory decision making (or of internal safety management, if a SWL owner/operator chooses to adopt internal safety goals). With respect to the proposed approach for ground-motion determination, it is only important that the definition of failure or unacceptable damage/contamination can be estimated using available methods of engineering analysis.

Once a probabilistic safety goal is established, and once the seismic hazard at a SWL site has been characterized and probabilistically quantified, the required seismic capacity of the SWL can be evaluated. This required capacity is determined through an iterative analysis procedure, whereby a trial seismic capacity is assumed, the risk is calculated and compared against the specific safety goal, and the trial capacity is appropriately adjusted until the safety target is achieved. The resulting required SWL capacity that just achieves the safety goal may be represented as a motion spectrum, a motion time history, or as a suite of motion spectra or time histories. Hence, once the required capacity is defined in this manner, essentially deterministic procedures can be used to perform the seismic evaluation or design. The engineer responsible for evaluation or design, therefore, need not implement a probabilistic analysis; rather, more familiar deterministic methods of analysis may be used.

The following sections provide a more complete outline of this approach for determining probabilistic risk-based motion input, including the roles of deterministic and probabilistic methods.

Performance Assessment Considerations

Seismic Risk-Assessment Framework. To determine risk-consistent, site-specific ground-motion input for seismic design or evaluation purposes, a probabilistic risk analysis framework is used. Such a framework is needed in order to compute failure probabilities, for comparison with safety goals. A general seismic risk-assessment framework can be expressed as follows:

$$\lambda_{State} = \Sigma_{(EQ\ Sources)}\ \nu_{Source} \left\{ \int_M \int_R P[State \,|\, M{=}m, R{=}r]\ f_M(m)\ f_R(r)\ dr\ dm \right\}_{Source}$$

Eq. (1)

where: λ_{State} is the annual frequency of a particular state of interest (e.g., liner breach, unacceptable environmental contamination, spectral acceleration exceeding 0.5g, etc.); a summation takes place for all earthquake sources (EQ Sources); ν_{Source} is the activity rate of a given seismic source (i.e., annual number of earthquakes which occur on the given source and which have magnitudes within the integration bounds over magnitude); variables M and R denote magnitude and distance, respectively; $f_M(m)$ and $f_R(r)$ are the probability density functions for magnitude and distance, respectively, for earthquakes occurring on the given source; and P[State|M=m,R=r] denotes the probability that the state of interest is realized, conditional upon occurrence of an earthquake (on the given source) of given magnitude at a given distance from the SWL site. Equation (1) is similar to the conventional integration framework for seismic hazard analysis, except a more general status probability is assessed within the integral, as opposed to a ground-motion exceedance probability (i.e.,

P[A>a|M=m,R=r]).   For purposes of numerical evaluation, and to assist with
conceptualization, Equation (1) can be re-expressed in the following discrete-integration
form:

$$\lambda_{State} = \Sigma_{i=1}^{N_S} \, \nu_i \, \Sigma_{j=1}^{N_M} \, \Sigma_{k=1}^{N_R} \, (P_{m_j})_i \, (P_{r_k})_i \, (P[State \,|\, m_j, r_k])_i$$

Eq. (2)

where: N_S, N_M and N_R, respectively, denote the number of seismic sources, number of
integration magnitudes, and number of integration distances; Pm_j is the probability of
magnitude for magnitude-integration interval j, evaluated from the magnitude probability
density function; and Pr_k is the probability of distance for the k-th distance-integration
interval. (Integration over distance is usually achieved by assigning equal weights to a
variety of locations of generated ruptures, whose size depends on magnitude, and then
evaluating the relevant distance from the site to each rupture location).

Probabilities of magnitudes and distances are evaluated internally in computer programs that
perform probabilistic seismic hazard analysis. These probabilities can be readily generated,
and they change only for different SWL sitings, not for different SWL designs at a given
site. The determination of values of P[State|m_j,r_k] -- which (depending upon the particular
states of interest) may require response analyses for each SWL design, and may require
ground-water transport analyses -- will likely become complicated (and expensive) unless
simplifying assumptions in the risk-assessment format are made. Following are discussed
detailed, conventional and recommended formats for implementing the seismic risk-
assessment framework.

Detail Seismic Risk-Assessment Format. As just alluded, the most general and accurate
evaluation of P[State|m_j,r_k] values, where acceleration time histories are used to characterize
the motion input, can become quite onerous. For example, we may suppose that the state
of interest is failure/breach of the SWL liner system. To evaluate P[Liner Breach|m_j,r_k], one
may collect several real records, or synthetically generated motions, that are applicable to
the given magnitude and distance, and are representative of expected site motions. Such a
collection of records would need to be obtained for each magnitude-distance pair considered
in seismic risk integration. For each record, a deterministic time-history analysis of SWL
response (for simulated model parameters) would be conducted, and an assessment of
maximum liner strain would be made. Based on a probability distribution of liner-material
resistances to strains, a probability of liner breach would be evaluated for the given
magnitude-/distance-dependent motion. A similar SWL response analysis and failure
evaluation would need to be conducted for each motion in the magnitude-/distance-
dependent suite (combined with a SWL model simulation). In turn, a similar suite of failure
analyses would need to be performed for all magnitude-distance pairs. Once all magnitude-
/distance-dependent failure probabilities had been evaluated, the integration of Equation 2
could be completed, producing a value for annual frequency of liner breach.

In this approach, if 50 motions and model-parameter combinations are generated for each
magnitude-distance pair, and 10 magnitude intervals and 20 distance intervals are used for
integration of risk results, then 10,000 failure analyses would need to be conducted. Clearly,

it would be impractical to perform such a large number of detailed response analyses just to evaluate a single SWL design! The use of simplified response models may make the approach feasible, but the accuracy of the results may very well suffer.

A more practical implementation of the detailed risk-assessment format, which still involves significant analytical effort, is to develop an expected failure-probability attenuation function P[State|m,r]=y(m,r) based on an appropriate set of combinations of ground-motion records and model-parameter simulations which are applicable for various magnitudes and distances (Sewell and Wu, 1994). A failure-probability attenuation function must be developed for each SWL configuration being evaluated and for each failure state of interest. For each motion record, a deterministic response analysis is performed, and a failure assessment is made, using the previously described procedure. The function y(m,r) is then fitted to this computed failure data, so as to define the expected failure probability in terms of magnitude and distance. For each magnitude-distance bin used in risk integration, the expected value of P[State|m_j,r_k] is determined by evaluating y(m,r) at M=m_j and R=r_k. Failure-probability attenuation functions generally die out at a much faster rate, with increasing distance and decreasing magnitude, than do ground-motion attenuation functions. (However, this would not be true for a seismically weak SWL that may be expected to fail for small-magnitude or distant earthquakes). Development of any given failure-probability attenuation function may require several hundreds of analyses, as opposed to the several thousands of analyses involved in the earlier approach.

Conventional Seismic Risk-Assessment Format. The more common, simplified approach to evaluating seismic failure probability requires a modification to the foregoing general risk-assessment framework. Hence, Equation 2 may be changed to the following form:

$$\lambda_{State} = \Sigma_{i=1}^{N_S} \ v_i \ \Sigma_{j=1}^{N_M} \ \Sigma_{k=1}^{N_R} \ (P_{m_j})_i \ (P_{r_k})_i \ \left\{ \Sigma_{l=1}^{N_X} \ (P[X \ge x_l | m_j, r_k])_i \times P[C_{State} = x_l] \right\}$$

$$\text{Eq. (3)}$$

where: X is a scalar measure of motion strength (e.g., spectral acceleration, inelastic spectral acceleration, PGA); C_{State} denotes the capacity of the SWL to resist the failure state, and is expressed only in terms of variable X (i.e., capacity no longer depends explicitly on magnitude and distance); and N_X is the number of integration points over motion amplitude.

Since P[C_{State}=x] is now independent of magnitude, distance, and seismic source, the following may be written:

$$\lambda_{State} = \Sigma_{l=1}^{N_X} \ P[C_{State} = x_l] \times \left\{ \Sigma_{i=1}^{N_S} \ v_i \ \Sigma_{j=1}^{N_M} \ \Sigma_{k=1}^{N_R} \ (P_{m_j})_i \ (P_{r_k})_i \ (P[X \ge x_l | m_j, r_k])_i \right\}$$

$$\text{Eq. (4)}$$

Equation 4 may be reduced further to obtain the more common expression of reliability evaluation:

$$\lambda_{State} = \Sigma_{l=1}^{N_X} P[C_{State} = x_l] \times \lambda[X \geq x_l]$$

$$= \int_X f_C(x) \, \acute{G}_X(x) \, dx$$

Eq. (5)

where: $\lambda[X \geq x_l]$ is the annual rate of occurrence of motion amplitudes exceeding the value x_l; $f_C(x)$ is the probability density function of SWL capacity evaluated at $X=x$; and $\acute{G}_X(x)$ is the complementary cumulative function of annual rate for motion-stength parameter X. $\lambda[X \geq x]$, which can alternatively be expressed as $\acute{G}_X(x)$, is the typical result of a seismic hazard analysis, and is usually conveyed as a seismic hazard curve.

Alternative ways of expressing Equation 5 are given as follows:

$$\lambda_{State} = \int_X F_C(x) f_X(x) \, dx$$

$$= \Sigma_{l=1}^{N_X} P[C_{State} \leq x_l] \times \lambda_{x_l}$$

$$= \Sigma_{l=1}^{N_X} P[State|x_l] \times \lambda_{x_l}$$

Eq. (6)

where: $F_C(x)$ is the probability distribution function of capacity, evaluated at $X=x$; $f_X(x)$ is the probability density function for seismic load parameter X; and λx_l is the annual probability of occurrence of motion amplitudes within the l-th amplitude interval for numerical integration.

In the above expression, $P[State|x]$ (equivalent to $P[C_{State} \leq x]$) is usually conveyed as a seismic fragility curve. The last form of Equation 6 represents the typical manner in which seismic failure probability is assessed; its evaluation requires that the seismic hazard curve be used to obtain motion-interval probabilities (which is simply done by taking differences in ordinates of the hazard curve at motion-interval end-points). Hence, fragility-curve ordinates are multiplied by annual probabilities of motion strength, and the resulting values summed, to obtain an estimate of seismic failure probability.

This conventional formulation for seismic risk assessment makes the following significant assumptions: (1) that the motion is meaningfully characterized, for purposes of failure assessment, by the single motion parameter X; and (2) that magnitude and distance effects, other than those that impact motion parameter X, are inconsequential to the failure assessment. These assumptions can create significant problems for the failure assessment of SWLs and of other structures which exhibit significant nonlinear response (unless the chosen parameter, X, adequately characterizes magnitude/distance effects on nonlinear response). Specifically, for a fixed motion strength, $X=x$, the damage to a SWL is likely to be greater from a distant, large-magnitude earthquake than from a close-in small-magnitude event; in addition, the significance of motion-to-motion variability in spectral shape (for given magnitude and distance), which has important implications for nonlinear response and failure assessment, is lost.

A seismic fragility curve may be evaluated based on the standard approach, which assesses the product of median safety factors and combines the variabilities of the various (lognormally distributed) safety factors. Alternatively, response analyses of the SWL can be conducted for various combinations (perhaps 50-100) of SWL model parameters and ground motions, to generate a fragility curve directly. In either case, an input spectral shape is needed, which suggests another problem with the conventional risk-assessment approach: What spectral shape for fragility evaluation should be chosen, and how should it be obtained? In response, use of a uniform hazard spectral shape had been initially proposed for seismic risk assessment. However, such a shape is a weighted composite of spectral shapes associated with various magnitudes and distances. The aggregate spectral shape may likely not be representative of an expected real earthquake motion spectrum. Hence, motion time histories generated to match or envelope a UHS, may produce unrealistic demands in SWL response analyses. Consequently, a UHS shape may not be well suited as a basis for SWL risk evaluation (see Sewell and Wu, 1994).

Recommended Seismic Risk-Assessment Format for SWL Evaluation. Recently, investigators have proposed that selection of a spectral shape (or a small suite of spectral shapes) for seismic fragility assessment should be based on magnitude-/distance-dependent considerations (see EPRI, 1994; and Sewell, 1989). If properly chosen, the magnitude-distance pair(s) used to derive the spectral shape(s) will produce input characteristics (and hence, response demands) that are better representative of expected real earthquake motions. Existing approaches have proposed obtaining the magnitude-distance pair(s) based on evaluation of expected magnitude and expected distance derived from contributions to the total annual frequency of exceedance of a specified motion amplitude of interest (e.g., 10^{-4} motion). Here, however, it is proposed that expected magnitude and expected distance be obtained from contributions to the total annual failure frequency. The logic behind this approach, is that spectral shapes (from magnitude-distance pairs) which do not result in SWL failure are not relevant to SWL fragility assessment. Notwithstanding this obvious observation, an apparent paradox involved in implementing this approach is that, in order to obtain contributions to failure frequency, one must have performed magnitude-/distance-dependent failure assessments; if such assessments have already been performed, the objective of implementing a simpler fragility analysis approach is defeated.

A valid, albeit approximate, way to avoid performing magnitude-/distance-dependent failure/fragility assessments, for use in this approach, is to initially assume or estimate an applicable fragility function. This fragility function is used to evaluate contributions to failure frequency coming from each magnitude-distance bin considered in seismic risk integration. These contributions are then used as weighting factors to evaluate mean magnitude and mean distance associated with failure. For the most general treatment, the fragility function may depend on magnitude and distance; for instance, the variability in capacity (characterized by the logarithmic standard deviation, β) will be lower for any given magnitude-distance bin, as compared to the variability defining a fragility function applicable to all magnitudes and distances. Research is currently being conducted to more precisely define how magnitude-/distance-dependent fragility functions differ from their conventional magnitude-/distance-independent counterpart form. This research will help enable a more refined assessment of magnitude-distance contributions to failure frequency. However, at the present time, a meaningful treatment is still achieved by use of a single fragility function to develop magnitude-/distance-dependent weighting factors for defining expected magnitude and expected distance. This approach produces results for expected magnitude and expected distance that are substantially different than those produced when weighting factors are based on contributions to annual frequency of exceedance of a specified motion amplitude.

A recommended seismic risk-assessment format for SWL evaluation has been developed based on the preceding approach for obtaining expected magnitude and expected distance. An outline of steps is as follows:

1. Develop (or assume) an initial magnitude-/distance-independent fragility function for the SWL.

2. Perform a seismic hazard analysis to evaluate probabilities of magnitudes and distances. For each magnitude and distance combination in the analysis, evaluate a failure frequency contribution as:

$$\lambda[State|m,r] = v_i \times P_{m_j} \times P_{r_k} \times \left\{ \Sigma_{l=1}^{N_X} P[State|X=x_l] \times P_{x_l} \right\}$$

Eq. (7)

where $P[State|X=x_l]$ denotes the discrete form of the fragility function obtained in Step 1, and Px_l is the probability of the motion amplitude being in the l-th amplitude-integration interval. Px_l is evaluated by integrating the magnitude-/distance-dependent motion probability density function between the l-th amplitude-interval endpoints.

3. Sum the failure frequency contributions over all distances, magnitudes, and seismic sources, to obtain a total failure frequency. Normalize all failure frequency contributions by the total failure frequency, to obtain magnitude-/distance-dependent weighting factors, w_{jk} (i.e., fractional contributions to failure frequency for each magnitude-distance bin).

4. Determine an expected/weighted magnitude as:

$$E[M] = \Sigma_{j=1}^{N_M} m_j \times \left\{ \Sigma_{k=1}^{N_R} w_{jk} \right\}$$

Eq. (8)

and determine an expected/weighted distance as:

$$E[R] = \Sigma_{k=1}^{N_R} \; r_k \; \times \; \left\{ \Sigma_{j=1}^{N_M} \; w_{jk} \right\}$$

Eq. (9)

5. Determine a spectral shape from a set of ground-motion relationships, evaluated for the expected magnitude and expected distance. These motion relationships should be consistent with those used in the seismic hazard analysis. Fragility evaluations are typically conducted assuming a median spectral shape. For design purposes, where (as opposed to the purpose of achieving the most realistic evaluation) it is desirable to consider a conservative spectral shape, use of the 84th-percentile spectral shape may be preferred.

6. Use the magnitude-/distance-dependent failure-related spectral shape as input for performing detailed response analyses of the SWL. A suite of acceleration time histories, with spectra matching this target spectral shape, should be generated (or obtained, if possible, from the database of applicable real records). Response analyses for these motions may serve as the basis for developing a refined estimate of the SWL seismic fragility. If this fragility is significantly different from that assumed in the analysis for obtaining expected magnitude and expected distance, then the foregoing procedures, starting with Step 2, should be repeated. (However, it is not necessary to repeat the seismic hazard analysis). Obtaining a valid fragility function, following this approach, should take no more than two iterations, given a reasonable starting fragility function.

This recommended format for seismic risk assessment of a SWL may be used to produce an annual frequency for liner breach, annual frequencies of unacceptable contamination at control locations, or other risk measures that may be of interest. The format is an improvement over conventional, simplified risk assessment procedures that do not make use of a realistic input spectral shape, and the format avoids the numerous analyses required for a more-rigorous full magnitude-/distance-dependent failure assessment.

Design/Evaluation Motion Selection and Characterization Process

At this point, it is necessary to demonstrate how a ground-motion input can be selected and characterized to be consistent with risk-based performance goals. Stated more explicitly, the objective is to obtain a ground motion that provides implicit assurance that seismic safety performance objectives will be satisfied, provided that the SWL can be demonstrated to possess (or be designed to possess) adequate capacity to resist the ground motion. Hence, as for any specified ground motion input, the engineer can use available deterministic methods, in the usual manner, in order to evaluate SWL response to risk-consistent motion input. If SWL response is shown to be adequate for the risk-consistent motion input, then correspondingly, it is shown that probabilistic safety goals have been met.

Procedure for Obtaining the Design/Evaluation Motion. The procedure for evaluating a risk-consistent ground-motion input is described as follows:

1.   Select the target safety goal to be met; e.g., mean frequency of seismic-induced liner breach not to exceed 5×10^{-4} per year.

2.   Perform a seismic hazard analysis for the site to obtain probabilities for magnitude-distance bins and to obtain a hazard curve, $\lambda[X \geq x]$. As discussed earlier, the motion parameter, X, used in developing the seismic hazard curve, should be a meaningful characterization of the potential of the motion to damage the SWL. Spectral acceleration, averaged over a frequency band from the expected initial frequency to the expected secant frequency, would be an appropriate parameter. Lacking an attenuation relationship for this parameter, spectral acceleration at the expected predominant response frequency of the SWL, would also be a valid parameter. (Inelastic spectral acceleration would be the best scalar parameter, but few applicable attenuation relationships are available, and the process for generating a motion time history to match a given inelastic spectrum is more complicated than that for matching a linear response spectrum).

3.   Select a trial median capacity for the SWL that is being evaluated (or being evaluated in design). Based on the assumption of a lognormal distribution of seismic capacity, construct the trial fragility curve according to the following equation:

$$F_C(x) = \Phi\left(\frac{\ln(x/\check{C})}{\beta}\right)$$

Eq. (10)

where $\Phi(\cdot)$ is the standardized cumulative normal distribution function; $\check{C}$ is the trial median seismic capacity, measured in terms of motion variable X; and β is the logarithmic standard deviation of capacity. For this analysis, it is sufficient to use a generic value for parameter β based on available fragility analyses results applicable to earth structures (e.g., dams, slopes, fills, etc.). Pending additional investigation, a value of $\beta=0.40$ may be selected as a generally reasonable choice.

4.   Integrate the site hazard curve and the trial fragility curve, according to Equation 6, to obtain a value of seismic risk (annual failure probability).

5.   Repeat Steps 2 to 4, as necessary, to obtain a median capacity, $\check{C}$, that results in a computed seismic risk that just meets the safety goal chosen in Step 1. Hence, if the computed risk from Step 4 is higher than the safety goal, increase the trial value of $\check{C}$; if the computed risk from Step 4 is less than the safety goal, decrease the trial value of $\check{C}$. Once the value of $\check{C}$ producing a seismic risk equal to the safety goal is obtained, the required seismic fragility curve of the SWL is fixed. It is then necessary to determine an associated spectral shape.

6.   A spectral shape is determined using the procedure described in the preceding section of this paper. However, because the desired fragility is already known, there will usually be no need to iterate using this procedure; one execution of the procedure should be sufficient.

Equation 7 should now be used to evaluate failure frequency contributions. The magnitude-distance probabilities have already been obtained in Step 2 above. Now, for each magnitude-distance bin, the conditional motion probability distribution, $P[X \geq x|m,r]$, is combined with the established risk-consistent fragility curve, to estimate magnitude-/distance-dependent contributions to failure frequency. These contributions are normalized by the total failure frequency to define weighting factors. The weighting factors are used in Equations 8 and 9 to determine expected magnitude and expected distance associated with SWL failure (at the safety-goal performance limit on failure frequency). The expected magnitude and expected distance are used to define an evaluation spectral shape, following Step 5 of the preceding section.

7.   The spectral shape determined from the magnitude-/distance-dependent assessment needs to be scaled such that the parameter X of the spectrum matches the required median capacity. For example, if X is based on spectral acceleration at 0.75 Hz, and the required median capacity is $\check{C}=0.20g$, then the spectral shape should be scaled to have a 0.75 Hz spectral acceleration of 0.20g. Because, however, the spectral shape has already been determined on the basis of the response spectrum expected to cause SWL failure, the scaling factor should not be too different from unity (i.e., it should not have to be scaled up or down dramatically).

Use of the Design/Evaluation Motion. Once the design/evaluation response spectrum has been obtained according to this procedure, a compatible time history or suite of time histories should be generated, as would be done for any evaluation motion. Response analyses of the SWL are then performed based on best-estimate material strengths, damping, geometry, etc., so that the evaluation analyses are best-estimate (i.e., median-centered) analyses. If the analyses indicate that the SWL can resist the evaluation/design motion without the failure state being reached, then the SWL has a median capacity in excess of the required median capacity, and hence, the SWL meets the desired performance safety goal. If, on the other hand, the failure state is reached or exceeded in the evaluation, the median capacity of the SWL is less than the required level, and the performance safety goal is not met.

For design, it is necessary to assume a SWL configuration to perform the median-centered evaluation. If the trial design is found not to meet the safety goal, then the properties of the landfill can be adjusted so that the evaluation produces satisfactory results (and hence, the safety goal is achieved). If the evaluation of the trial design reveals that the design is more than satisfactory, then (if greater economy can be obtained) adjustments can be made to lower the capacity of the landfill. In this way, an iterative design procedure, that results in acceptable SWL performance, is implemented.

Role of Deterministic and Probabilistic Methods of Analysis

The reader is encouraged to consider the following perspective: Probabilistic methods do not replace deterministic methods, nor do they replace engineering judgment. Probabilistic methods are simply tools that, like deterministic methods, an engineer is better off with, than without.

These statements are especially applicable to the seismic design or evaluation of SWLs, where deterministic engineering analyses are obviously needed, and yet, having a framework capable of meaningfully addressing uncertain material properties, uncertain SWL geometry, variable load conditions, and the like, is also a prime concern.

Development of the proposed approach to SWL evaluation has been undertaken to clarify and substantially separate the roles of deterministic and probabilistic analysis. All of the probabilistic analysis is here embodied in the design/evaluation ground motion. An engineer familiar with probabilistic methods can implement the approach to obtain the risk-based design/evaluation ground motion. Then, an engineer familiar with deterministic design/evaluation procedures can take the design/evaluation motion and check/insure that the SWL responds adequately (e.g., does not fail) when subjected to that motion.

The approach does not require that the design/evaluation engineer implement a risk analysis, nor does it require that the probabilistic engineer conduct detailed response evaluations. In this way, both engineers can concentrate on their respective expertise. In addition, the approach makes obvious that deterministic methods are not replaced by probabilistic methods; the approach retains clear prominence of deterministic methods in the design/evaluation process. Yet, the prominence of probabilistic methods in linking the design/evaluation process to safety criteria (expressed in terms of acceptable risk) is also made clearly evident. Thus, both types of methods should be viewed as useful tools important to SWL design and evaluation.

Risk-Based Performance Regulation

Solid waste landfills are regulated by the Environmental Protection Agency (EPA) of the federal government, by state environmental agencies, as well as by local governmental agencies. At present, regulatory criteria for seismic behavior of SWLs are not based on risk-limiting safety goals. EPA criteria for municipal solid waste landfills discuss seismic siting provisions for SWLs, which define a seismic impact zone as any region where the 2,373-yr ground motion exceeds 0.10g (U.S. EPA, 1991). Seismic design for SWLs located in such zones is based on the 2,373-yr ground motion or on the maximum expected motion. State of California regulations specify that a SWL must be designed to withstand the maximum probable or maximum credible earthquake, without damage to the foundation or to structures which control leachate, surface drainage, erosion, or gas (California Code of Regulations, 1992 and 1989).

These regulations are not sufficiently prescriptive so as to define a uniform, risk-limiting standard of safety for SWLs. More precise definitions of failure, and of failure frequency, need to be specified. In addition, the criteria do not explicitly specify a level of safety against unacceptable ground-water contamination.

The level of safety implied by the current criteria will vary from site to site. It is, however, possible to obtain a rough bound of the general level of safety that is implied by these criteria. The following equation represents a simplified analytical form of seismic failure-frequency assessment:

$$\overline{\lambda}_{Failure} = \overline{H}(\check{C}) \exp\left(\frac{(K_H \beta)^2}{2} \right)$$

Eq. (11)

where: $\lambda_{Failure}$ is the mean annual frequency of SWL failure (or damage); $H(\check{C})$ is the mean hazard evaluated at the median capacity of the SWL to resist failure (or damage); and K_H is the slope of the seismic hazard curve in logarithmic scale. For a 2,373-yr design ground motion (i.e., mean annual hazard of 0.00042 at the median capacity to resist damage), a typical value of K_H=3, and a representative value of β=0.4, a mean frequency of SWL damage is estimated to be about 9×10^{-4} (i.e., 10^{-3} level). There is an unknown margin of safety implied by the use of "damage" as the failure state of interest, as opposed to specifying unacceptable performance based on liner breach. Hence, the frequency of liner breach implied by the current regulatory criteria will be less than the level of 9×10^{-4}/yr; correspondingly, the frequency of unacceptable ground-water contamination at any control location will also be less than this level. If a conditional factor of safety of 1.2 against liner breach, given SWL damage, is assumed, then the median capacity would increase by 20%, the mean hazard at this increased capacity would decrease by about 40%, resulting in a mean frequency of liner breach of about 5×10^{-4}. If a conditional factor of safety of 1.5 is assumed, then the mean frequency of liner breach would be about 3×10^{-4}. These numbers are useful generic values for regulators and SWL safety managers to keep in mind when addressing risk concerns. A site-specific risk assessment would generally be required to assess the safety of a given design. It is proposed here, however, that a preferable alternative (that better correlates design with safety) is to establish design/evaluation motions on a risk-consistent basis, by following the guidance outlined earlier in this paper.

Summary and Conclusions

A workshop held during August of 1993 in southern California has identified research priorities for seismic design of solid waste landfills (Martin and Kavazanjian, 1994). These research priorities are still valid today; the need for more formal procedures for the determination and use of seismic risk results is one such identified priority.

This paper has attempted to provide some initial insights and guidance on the use (and usefulness) of seismic risk assessment in the seismic design of solid waste landfills. The general methods described in this paper have already been successfully applied in establishing risk-consistent design/evaluation demands for a variety of important engineering facilities (e.g., see Sewell, 1994; Sewell et al., 1994, 1993, 1990; Kennedy, 1992; and Kennedy et al., 1990). The procedures outlined in this paper enable an engineer to perform a deterministic design, with respect to a probabilistically derived design/evaluation motion, to insure that explicit safety-performance goals are satisfied. This paper has also identified, or alluded to, issues that still need to be addressed in seismic risk assessment of solid waste landfills. Additional, future research will be highly significant to the resolution of these concerns.

References

California Code of Regulations, Title 23. California Water Regulations, Chapter 15. "Discharges of Waste to Land," 1992.

California Code of Regulations, Title 14. California Waste Management Board, Chapter 3. "Minimal Standards for Solid Waste Handling and Disposal," 1989.

Electric Power Research Institute (EPRI). *Seismic Hazards/Design Parameters for a Nuclear Waste Repository.* EPRI TR-104233, Electric Power Research Institute. October 1994.

Martin, G.R., and E. Kavazanjian, Jr. (Organizers). *Proceedings of a Workshop on Research Priorities for Seismic Design of Solid Waste Landfills,* August 26-27 1993, University of Southern California. Sponsored by the National Science Foundation. Proceedings issue date: March 1994.

Mitchell, R.A. and J.K. Mitchell. "Stability Evaluation of Waste Landfills." In: *Proceedings of a Workshop on Research Priorities for Seismic Design of Solid Waste Landfills,* August 26-27 1993, University of Southern California. Sponsored by the National Science Foundation. Proceedings issue date: March 1994.

Kennedy, R.P., et al. "Performance goal based seismic design criteria for high level waste repository facilities," In Proceedings: *ASCE Symposium on Dynamic Analysis and Design Considerations for High-Level Waste Repositories,* San Francisco, CA. August, 1992.

Kennedy, R.P., et al. *Design and Evaluation Guidelines for Department of Energy Facilities Subject to Natural Phenomena Hazards,* Report UCRL-15910. Office of Safety Appraisals, U.S. Dept. Of Energy. 1990.

Kennedy, R.P., et al. *Engineering Characterization of Ground Motion - Task 1: Effects of characteristics of free-field motion on structural response,* NUREG/CR-3805, Vol. 1. February 1984.

Repetto, P.C., J.D. Bray, R.J. Byrne, and A.J. Aguello. "Seismic Analysis of LandFills." In: *Proceedings of a Workshop on Research Priorities for Seismic Design of Solid Waste Landfills,* August 26-27 1993, University of Southern California. Sponsored by the National Science Foundation. Proceedings issue date: March 1994.

Seed, R.B. and R. Bonaparte. "Seismic Analysis and Design of Lined Waste Fills: Current Practice." In: *Proceedings of a Workshop on Research Priorities for Seismic Design of Solid Waste Landfills,* August 26-27 1993, University of Southern California. Sponsored by the National Science Foundation. Proceedings issue date: March 1994.

Sewell, R.T. "Methods to use performance goals and seismic hazard results to determine seismic design parameters for high-level waste repositories," *EPRI TR-104233* (Section 4). October 1994.

Sewell, R.T. and S.C. Wu. *Impact of Ground-Motion Characterization on Seismic Evaluation Studies*. Nuclear Regulatory Commission, January 1994.

Sewell, R.T., R.K. McGuire, and G.R. Toro. *Use of Probabilistic Seismic Hazard Results: General Decision Making, the Charleston Earthquake Issue, and Severe Accident Evaluations*. EPRI TR-103126, Electric Power Research Institute, October 1993.

Sewell, R.T., T.F. O'Hara, C.A. Cornell, and J.C. Stepp. "Selection of review method and ground-motion input for assessing nuclear power plant resistance to potential severe seismic accidents. In Proceedings: *Third Symposium on Current Issues Related to Nuclear Power Plant Structures, Equipment, and Piping*. North Carolina State University, December 1990.

Sewell, R.T., et al. *Impact of Ground Motion Characterization on Conservatism and Variability in Seismic Risk Estimates*, NUREG/CR Report. 1989.

Sewell, R.T., and C.A. Cornell. *Nonlinear-Response-Based Measures of the Damage Potential of Earthquake Ground Motion and Their Use in Seismic Hazard Analysis*. Tech Rept. EPRI RP2556-8, Electric Power Research Institute, October 1988.

U.S. Environmental Protection Agency (EPA). "EPA Criteria for Municipal Solid Waste Landfills," 40 CFR 258; 56 FR 51016, *Code of Federal Regulations*. October 9, 1991.

U.S. Nuclear Regulatory Commission (USNRC). "Safety Goals for the Operations of Nuclear Power Plants; Policy Statement." *Federal Register* - 10 CFR Part 50, Vol. 51, No. 149. August 4, 1986.

GEOTECHNICAL CONSIDERATIONS IN THE SEISMIC RESPONSE EVALUATION OF MUNICIPAL SOLID WASTE LANDFILLS

Sukhmander Singh[1], M. ASCE
Joseph Sun[2], M. ASCE

Abstract

This paper discusses the application of seismic design procedures based on conventional geotechnical earthquake engineering practice in the seismic response evaluation of Municipal Solid Waste Landfills (MSWL). The authors looked critically at geotechnical considerations that are important in understanding the behavior of MSWLs. The paper also addresses concerns that are unique to dynamic material properties and seismic response of MSWLs.

Introduction

Seismic response analyses of MSWLs are typically carried out using procedures developed for soil slopes or earth dams. Accordingly, assessment of landfill stability under earthquake loading conditions is commonly performed using the sliding block deformation analysis method developed by Newmark (1965). Typical deformation analyses employ a combination of one-dimensional seismic response analyses, e.g. SHAKE and Newmark sliding block deformation procedures, to evaluate permanent deformations which are assumed to take place along a sliding surface. An accurate estimation of the material properties, especially the dynamic strength properties, is essential for those analyses. Refuse is, however, a highly heterogeneous material, and its dynamic strength properties are yet to be established through a reasonable set of data measured or observed. There are other aspects of the MSWL that differ from a well-compacted homogeneous earth dam---for example, the relatively light weight of refuse, and the placement of layers of soils and liner systems. Due to these differences, there are concerns regarding the applicability of

[1]Professor and Chairman, Dept. of Civil Engineering, Santa Clara University, Santa Clara, California 95053

[2]Senior Project Engineer, Woodward-Clyde Consultants, Oakland, California, 94607

the conventional geotechnical earthquake engineering procedures to landfill refuse slopes.

Questions have also been raised about the applicability of Mohr Coulomb's Theory to the characterization of the shear strengths of refuse (Singh and Murphy, 1990; Mitchell and Mitchell, 1992). The validity of limit equilibrium analysis to evaluate the stability of refuse slopes has been of concern due to the incompatibility of strains that cause failure in soil with those that cause failure in refuse material (Singh and Murphy, 1990; Mitchell and Mitchell, 1992; Fassett, et al., 1994). Current practice in the seismic analysis and design of solid waste landfills was addressed by Seed and Bonaparte (1992). Repetto et al. (1993) examined the application of wave propagation analysis and selection of seismic design parameters for refuse landfills. In a state-of-the-art paper, Anderson and Kavazanjian (1995) discussed the performance of landfills under seismic loading and critically reviewed the seismic response analysis of MSWLs. Mitchell et al. (1995) described the nature and consequences of material interactions, unique to MSWLs, that must be considered in the analysis and design of solid waste landfills. These studies have suggested that the conventional geotechnical seismic design procedures must evolve to adequately address the unique concerns of MSWLs. This paper critically examines the geotechnical considerations that apply to the evaluation of refuse behavior, and the significant concerns in the seismic response evaluation of MSWLs.

Characterization of the Shear Strength of Municipal Solid Waste (MSW)

Fassett, et al. (1994) presented an excellent summary and analyses of MSW strength properties reported in several papers to date (1993-94) and brought into focus the limitations of existing approaches to characterize the shear strength properties of MSW. Characterization of shear strength of refuse material has been attempted essentially in two ways. Singh and Murphy (1990) summarized existing data from laboratory tests, from back-calculations, and from in-situ testing, and recommended a range of strength parameters expressed in pairs of c and ϕ for the MSW. Howland and Landva (1992) used an alternate method and expressed MSW strength in terms of mobilized shear strength and normal stress. Howland and Landva considered the strength of MSW to be primarily frictional in nature. In terms of Mohr-Coulomb parameters, the relationship between shear strength and normal stress developed by Howland and Landva gave a c equal to 10 kPa and ϕ equal to 23 degrees as a lower bound to their data. Kavazanjian, et al. (1995) used an approach similar to that of Howland and Landva, except that in their analysis, Kavazanjian, et al. relied more on data based on back calculations from case histories and in-situ testing than they did on laboratory testing data. Kavazanjian, et al adopted a bilinear representation of the MSW shear strength using Mohr-Coulomb parameters. They suggested that at normal stress below 30 kPa, the Mohr-Coulomb parameters were $\phi = 0$ and $c = 24$ kPa, and above normal stress of 30 kPa, the parameters were $c = 0$ and $\phi = 33°$.

All of the foregoing investigators expressed concerns about the compressibility and strain compatibility as it relates to the refuse shear strength when used in a limit equilibrium slope stability analysis. The following section critically examines the assumptions made in characterizing the MSW strength.

Characterization of MSW strength in terms of c and ϕ is based on the premises that there are soil layers in an MSWL and that the strength of MSW increases as the confining pressures increase. The presence of cohesion as it exists in a mineral soil in the form of grain-to-grain attraction may not be present in the MSW. However, since vertical cuts made in MSW have been reported to stand safely (Purcell, Rhoades and Associates, 1987 and Earth Technology, 1988), existence of an apparent cohesion cannot be ruled out. Mitchell and Mitchell (1992) argue that because of the interlocking and overlapping of refuse constituents, MSW exhibits an apparent cohesion. Most laboratory test investigators, however, have classified MSW as primarily a frictional material. It can be argued that the fibrous nature of some of the constituents of MSW can manifest itself more predominantly in the form of apparent cohesion in a prototype than in a laboratory test run on a finite size sample of MSW.

During shear, the strength mobilization rates for MSW and soils are quite different. The MSW behaves as a strain hardening material, i.e., even at large shear strain, it will continue to mobilize additional shear stress without exhibiting a levelling off or drop in shear stress or developing a failure plane typical of soils. In a limit equilibrium analysis, such as used in slope stability analysis, the basic assumption is that the peak strength mobilization occurs at the same time all along the failure surface. Because of the incompatibility of strains at which peak strength is mobilized in MSW, in soils, and along liner interfaces, the limit equilibrium analyses need to use reduced shear strength of MSW or the residual strength of soil and the interfaces (Mitchell et al. 1995).

Characterization of MSW strength primarily as a frictional material may lead to an overestimation of strength under high confining pressures; for example, because of the non-granular, fibrous, and oddly shaped nature of the constituents of MSW, the possibility of significant arching action is real. Furthermore, the existence of relatively large differential settlements which are common within MSWL, bridging over or arching actions can be expected. Accordingly, estimation of shear strength based directly on an increase in overburden pressure implies that the strength continues to increase linearly with increasing depth of the landfill. This is not the case in cohesionless soils as shown by Vesic (1967, 1970) and Meyerhof (1976). Based on their studies the concept of critical depth, below which vertical effective pressure does not contribute to the shear strength, is routinely used to estimate the unit skin friction for piles in sands.

Singh and Murphy (1990) indicated maximum shear strength of MSW, expressed as cohesion intercept, is on the order of 110 kN/m². For deeper MSW, we hypothesize that the lower bound for maximum shear strength will likely be 250 kN/m^2. On these bases, the authors developed a strength-versus depth plot for MSW as shown in Figure 1. For comparison purposes, Figure 1 also shows MSW strength characteristics developed by Howland and Landva (1992) and Kavazanjian et al. (1995).

Based on the limiting shear strength value of 250 kN/m^2, this would correspond to a critical depth of 55 meters. It is interesting to note that in a recent study of the variation of refuse unit weight versus depth, Kavazanjian et al. (1995) have shown that the unit weight increases with depth, but below about 50 m depth, the unit weight remains mostly constant (Figure 2). Whereas there are other factors such as settlement and composition of the refuse, which influence the increase in unit weight with depth, compression under the overburden is one of the significant factors. Accordingly, it appears that the influence of overburden in compressing the refuse material below about a 50-meter depth reaches to a minimal level. Conduto and Huitric (1990) investigated movements within a landfill by installing instruments inside vertical borings drilled through a landfill in order to monitor both vertical and horizontal movements at various depths. Data collected over a period of 2 years suggested that vertical strain rates are independent of depth. Again the influence of vertical overburden appears to be not significant. In view of the foregoing discussions, the authors have shown the probable variation of shear strength with depth in Figure 1. However, more data is needed to verify the curve for depths greater than 50 meters.

Finally, there is the age factor. Arching action may dominate in younger landfills, whereas differential settlements may dominate in older landfills due to decomposition of MSW constituents with time. How to account for these in characterizing strength of MSW is an important question. Can a small or a finite size sample of refuse represent the phenomena of arching and differential settlements in MSWL? Landva and Clark (1990) noted that a laboratory shear test cannot detect the existence of cavities or weak zones formed by substantial local decompositions within the fill. Therefore, field measurements of density and shear strength supplement with back calculation of well documented case histories appear to be the most logical mean of obtaining MSW strength information. These are some of the important questions which should be kept in view when applying geotechnical considerations in the characterization of shear strength of MSW.

Dynamic Stress-Strain Properties of MSW

A rational development of the dynamic strength parameters of a municipal solid waste material has been very limited. This is because of: 1) the highly heterogeneous nature of refuse material, 2) its changing characteristics with time by decomposition, and 3) the difficulties in sampling and testing representative samples in the laboratory. Because of the absence of laboratory data, early investigators assumed solid waste material to behave similarly to peat, clay, or a combination of

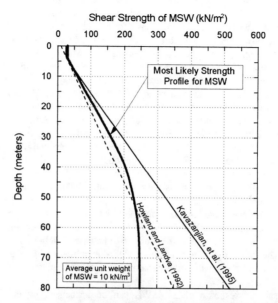

Figure 1   Shear Strength Variation with Depth for MSW

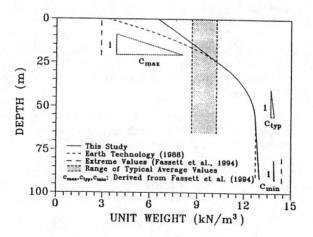

Figure 2   Unit Weight Profile for MSW (Kavazanjian et al., 1995)

peat and clay e.g., Singh and Murphy (1990), Earth Technology (1988), and Sharma and Goyal (1991). Downhole shear wave velocity data measured at Richmond and Redwood landfills was used to provide a basis for modulus curve at low strain levels (Singh and Murphy, 1990). Stewart, et al. (1994) reported that the MSW modulus reduction and damping curves developed by Singh and Murphy (1990), which was derived from peat and clay curves (Seed and Idriss, 1970; Sun et al, 1988), gave good agreement between observed and predicted response at the top of an instrumented landfill. This is confirmed by Kavazanjian and Matasovic (1995) modulus reduction and damping curves which is based on back calculations and best fit parameters for the recorded data at OII Landfill in Los Angeles during the Northridge Earthquake as shown in Figure 3. However, much remains to be learned about dynamic properties of refuse at higher strain levels and this subject is debatable because more recent field investigations at OII landfill indicate that the waste may be more elastic over a wider strain range than those shown in Figure 3. It is not clear if this phenomenon also applies to other landfills. Until large-sized representative samples of refuse can be tested, and in situ tests are developed, or, as pointed out by Anderson and Kavazanjian (1995) additional data from more intense shaking are obtained, the shape of the modulus reduction and damping curves at strains greater than 2×10^{-2} percent will continue to be based solely upon engineering judgement. The following section presents a critical examination of the geotechnical considerations used in estimating the dynamic strength properties of MSW.

The seismic response of a MSWL can be influenced by the following considerations unique to MSW:

- The MSW material appears to have elastic properties at low strain levels

- The very low shear strengths along the interfaces of a modern composite liner system can act as a base isolation system

- Since the MSW decomposes with time, the dynamic strength properties will vary with time. For a younger MSWL, where the interlocking and /or overlapping dominate, the response can be influenced by both material and system damping

- The elastic properties may dominate at low strain levels. Accordingly, the damping of MSW at very low strain levels may not exist or it may be negligible due to the elastic properties of the MSW material.

The influence of low shear strength interfaces has been investigated by Kavazanjian and Matasovic (1995). The results indicated a reduction in seismic motions due to low strength interfaces. In terms of displacements along the interfaces, the results can be unfavorable.

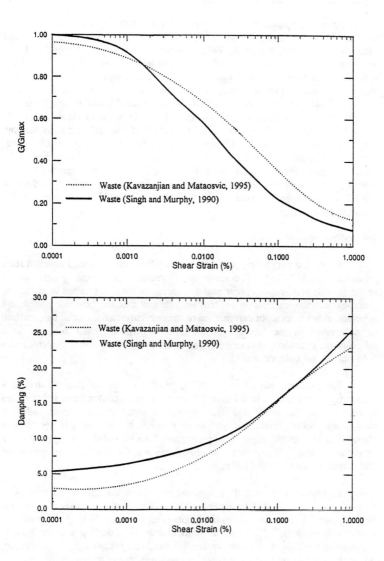

Figure 3 Shear Modulus Reduction and Damping Curves

The Seismic Response Evaluation of MSWL

Seismic response of MSWL has been evaluated using one-dimensional models (Earth Technology, 1988; Singh and Murphy, 1990; Kavazanjian and Matasovic, 1995; Repetto, et al., 1993; and Bray, et al., 1995). Equivalent linear and non-linear procedures have been employed in modeling the one-dimensional response. The municipal solid waste landfill geometry, especially when it is fitted with low strength interfaces, may warrant two- or three-dimensional analyses. Failure of the Kellement-Hills landfill during filling brought into focus the importance of two- or three-dimensional analysis. Anderson and Kavazanjian (1995) suggest that whenever three-dimensional effects are of concern for gravity loading, they must be considered for seismic loading. However, analyses using two-dimensional models for seismic response have not been extensively reported in the literature, even though these have been employed or considered by some investigators. In the meantime, the following section presents spectral analysis of MSWL to further the understanding of the behavior of landfills during earthquake loading conditions.

Spectral Analysis of MSWL

As pointed out earlier, one of the major differences between municipal solid waste landfills (MSWLs) and conventional soil structures is the heterogeneous nature of the landfill material. Conventional soil mechanics was developed based on the premises that the material is uniform enough and that the behavior of the soil structure, either in terms of deformation or strength characteristics, can be simulated or determined in the laboratory. In other words, the soil's behavior in a microcosmic situation (the laboratory) can be used to predict the macrocosmic soil's behavior (in the field, on a site).

The materials placed in a MSWL usually cannot be characterized in a microcosmic situation. The fill material can range from conventional soil particles (used to form the liners and soil covers) to anything imaginable, including construction debris, metal, and perishable material of any size. If one tries to characterize the landfill engineering properties of the fill based on samples that can fit in the soil testing apparatus or even in a large box, the results can be misleading and inappropriate for landfill design.

To characterize landfill engineering properties, the appropriate way is to perform large-scale field tests supplemented with back analyses from past landfill performances. In other words, landfill engineering characteristics are determined in a macrocosmic sense. Nothing can be more true when dealing with the seismic response of landfills. Unfortunately, landfill monitoring is not a common practice; not even in the environmentally conscious and seismically active California. The only MSWL carefully monitored with geotechnical instruments in the U.S. is the

Operating Industries, Inc. (OII) Landfill in southern California. OII Landfill is located about 10 miles east of Los Angeles, within the city of Monterey Park. The landfill contains a 45-acre north parcel and a 145-acre south parcel, separated by the Pomona Freeway, which runs in an east-west direction. The top and bottom of the landfill are at approximate elevations of 630 and 380 ft, respectively. The OII Landfill is well instrumented with surface settlement monuments, slope inclinometers, extensiometers, piezometers, and most importantly, a pair of seismographs. The seismographs are located on the eastern slope of the south parcel. One is placed on top of the fill, approximately 100 feet from the edge of the slope, and the other one is placed at the toe of the fill on native firm soil. Since the installation of these accelerograms in 1988, over 20 earthquakes have been recorded, including the 1988 Pasadena earthquake (M=5), 1992 Landers earthquake (M=7.3), Big Bear earthquake (M=6.4), and Joshua Tree earthquake (M=6.1), and the 1994 Northridge earthquake (M=6.7). With the except of 1992 Joshua Tree earthquake, which recorded less than 0.01g at the base, the remaining 4 earthquakes were used to develop the normalized response spectrum discussed below. The largest accelerations recorded at the site were those triggered during the 1994 Northridge earthquake with base accelerations of 0.25g (Longitudinal) and 0.23g (Transverse), and crest accelerations of 0.26g (Longitudinal) and 0.21g (Transverse).

In order to understand the overall response of MSWL and correlate it with the widely used site dependent spectra, we developed the normalized response spectra for the 36 motions recorded on top of the landfill, as shown on Figure 4. The dashed line represents the average normalized response spectrum for all motions recorded. The solid line represents the average spectrum for earthquakes with magnitude greater than 5. Several interesting observations can be made from this figure. First, the long period motions are richer for earthquakes with magnitude greater than 5. Second, the site period at about 1.2 seconds clearly demonstrates the response spectrum at long periods. Third, the site period does not seem to change significantly from low magnitude earthquakes to high magnitude earthquakes. The following paragraphs will address each of these three observations and their implications for design issues.

The richness in long-period motions shown in Figure 4 can be attributed to two possible sources: 1) input motions carry more long-period motions for large-magnitude earthquakes, and 2) larger magnitude earthquakes usually are longer and allow the landfill to have more time to develop its fundamental mode of vibration. The first source is commonly known in the field of seismology and widely accepted among practicing engineers. The second source is demonstrated in Figure 4, which shows that the spectral amplitude difference between the two normalized spectra is not uniform. That should be the case if the difference in long-period input is the only source. As a matter of fact, the maximum difference occurs near the site period; this further demonstrates that site response plays an important role in contributing to the long-period motions at the OII site.

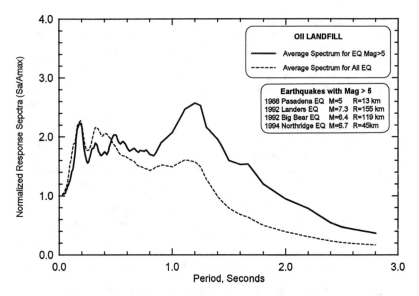

Figure 4   Normalized Response Spectra for Recorded Motions at Top of OII

The effect of site amplification for sites underlain by clays has long been recognized (Seed, et al., 1974; Seed and Sun, 1989). The 1985 Mexico City earthquake is probably the best example of this phenomenon. The dominance of the site period on the response spectrum for the OII Landfill is much too strong to be overlooked, especially for earthquakes with magnitudes greater than 5. Figure 5 shows the comparison between the OII normalized response spectra and those published by Seed et al. (1974) for various site conditions. Based on this plot, it can clearly be seen that the response of a landfill resembles a clay site even though in slope stability analyses it has conventionally been treated as a cohesionless material. Figure 6 compares the two OII normalized spectra with the Uniform Building Code (UBC) site-dependent spectra. The average response spectrum for earthquake magnitudes greater than 5 clearly demonstrates that the OII Landfill responded more like an S3 site than an S2 site in major earthquakes. On the above basis, we believe that for the present time and without site specific investigations, all landfill sites should be designed with at least UBC S3 site conditions and a site-specific investigation will be appropriate for detailed design of structures built on landfills.

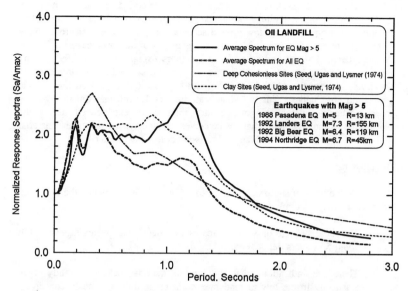

Figure 5   Comparison of OII Spectra with Site Dependent Spectra

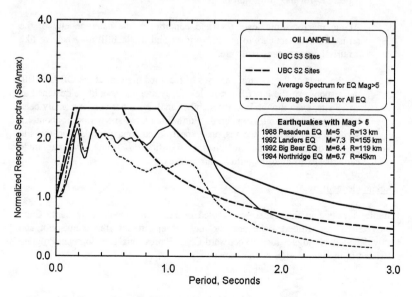

Figure 6   Comparison of OII Landfill Spectra and UBC Response Spectra

The third observation made here is that the site period does not seem to vary far between the high magnitude earthquakes with high ground acceleration and low magnitude earthquakes with low ground accelerations. It can be seen from Figure 4 that the site period stayed at 1.2 seconds regardless of the earthquake magnitude. A more detailed study performed by the authors (Singh and Sun, 1995), using Fourier spectral analyses to determine the site period from the 1994 Northridge main shock and aftershocks which have 0.25g and less than 0.05g, showed that the site period did not change. This would indicate that the landfill material may behave more elastically in the low-to-moderate strain range experienced during the Northridge earthquake than what we perceived previously. However, the material non-linearity of the fill at high strain levels is yet to be determined.

Conclusions:

On the basis of studies and discussions presented in this paper, the following conclusions may be drawn.

1.   The shear strength of MSW increases with depth but appears to level off at about 55 meters depth (Figure 1).

2.   There are indications that the landfill material behaves slightly more elastically in the low to moderate strain range than previously conceived. Accordingly, damping in the material is slightly less than previously thought, especially for the low strain ranges.

3.   At the macrocosmic level (in field conditions), the heterogeneity of the refuse material appears to minimize itself and the landfill may behave like a fairly homogeneous structure.

4.   On the basis of the spectral analyses of recorded motions at the OII Landfill, the response of the landfill resembles a clay site and should be designed as an S3 site, even though in slope stability analyses it has conventionally been treated as a cohesionless material. Furthermore, the site period at about 1.2 seconds demonstrates the response spectrum at long periods; and the site period does not seem to change significantly from low magnitude earthquakes to high magnitude earthquakes.

Acknowledgements

Studies reported in this paper were based on research supported by Santa Clara University. Dr. Robert Pyke and Dr. Lelio Mejia review the manuscript and provided valuable suggestions. Woodward-Clyde Professional Development Program assisted in the preparation of this paper.

References

Anderson, D. G., and Kavazanjian, E., (1995), "Performance of Landfills Under Seismic Loading," Invited state-of-the-art paper, proceedings of the Third International Conference on Recent Advances in Geotechnical Earthquake Engineering and Soil Dynamics, University of Missouri at Rolla, April 2-7.

Bray, J. D., Augello, A. J., Repetto, P. C., Leonards, G. A., and Byrne, R. J. (1995), "Seismic Analytical Procedures for Solid Waste Landfills," submitted to the Journal of Geotechnical Engineering ASCE, Vol. 121, No. 2, February, pp. 136-151.

Coduto, D. P. and Huitric, R. (1990) "Monitoring Landfill Movements Using Prceise Instruments," Geotechnics of Waste-Theory and Practice, ASTM, STP 1070, pp. 358-370.

Earth Technology (1988). "In-Place Stability of Landfill Slopes, Puente Hills Landfill, Los Angeles, California," Report No. 88-614-1, prepared for the Sanitation Districts of Los Angeles County, The Earth Technology Corp., Long Beach, CA.

Fassett, J. B., Leonards, G. A., and Repetto, P. C., (1994), "Geotechnical Properties of Municipal solid Wastes and Their Use in Landfill Design," Proc. WasteTech '94 - Landfill Technology Conference, National Solid Waste Management Association, Charleston, South Carolina, p. 32.

Fassett, J. B., Leonards, G. A., and Repetto, P. C. (1994) "Geotechnical Properties of Municipal Solid Wastes and Their Use in Landfill Design," Proc. WasteTech '94, Charleston, SC, National Solid Waste Management Association, Washington, D.C.

Howland, J. D. and Landva, A. O. (1992), "Stability Analysis of a Municipal Solid Waste Landfill." Proceedings, Stability and Performance of Slopes and Embankments - II, Vol. 2, Geotechnical Special Publication No. 31, ASCE, New York, NY.

Kavazanjian, E., Jr., and Matasović, N. (1995), "Seismic Analysis of Solid Waste Landfills" Accepted for publication, Proceedings, GeoEnvironment 2000, Geotechnical Special Publication, ASCE, New York, NY.

Kavazanjian, E., Jr., Matasović, N., bonaparte, R., and Schmertmann, G. R. (1995) "Evaluation of MSW Properties for Seismic Analysis" Accepted for publication, Proc. GeoEnvironment 2000, Special Geotechnical Publication, ASCE.

Meyerhof, G. G. (1976) "Bearing Capacity and Settlement of Pile Foundations," Journal of Geotechnical Engineering Division,ASCE, 102, No. GT3, March.

Mitchell, R. A., and Mitchell, J. K., (1992) "Stability Evaluation of Waste Landfills," Proceedings of ASCE Specialty Conference on Stability and Performance of Slopes and Embankmcnts - II, Berkeley, California, June 28 - July 1.

Newmark N. M., (1965) "Effects of Earthquakes on Dams and Embankments," The Rankin Lecture. Geotechnique. Vol. 29, No. 3, pp. 215-263.

Mitchell, J.K., Bray, J.D. and Mitchell, R.A., (1995) "Material Interactions in Solid Waste Landfills", Proceedings, GeoEnvironment 2000, ASCE, Geotechnical Special Publication No. 46, Vol. 1, pp. 568-590.

Newmark N. M., (1965) "Effects of Earthquakes on Dams and Embankments," The Rankin Lecture. Geotechnique. Vol. 15, No. 2, pp. 139-160.

Purcell, Rhoades and Associates (1988), "Stability Analysis - Static and Dynamic Loading Conditions, Sunshine Canyon Extension Site, Sylmer, California", Report presented to BFI Industries of California, April.

Repetto, P. C., Bray, J. D., Byrne, R. J. and Agruello, A. V., 1993) "Seismic Analysis of Landfills," *Proceedings of the 13th Central Pennsylvania Geotechnical Seminar,* April 12-14.

Seed, H. Bolton, Ugas, and Lysmer, John (1974) "Site-Dependent Spectra for Earthquake-Resistant Design," Earthquake Engineering Research Center Report No. EERC 74-12, University of California, Berkeley, November.

Seed, H.B. and Idriss I.M. (1970) "Analyses of Ground Motions at Union Bay, Seattle During Earthquakes and Distance Nuclear Blasts", Bulletin of the Seismological Society of America, Vol. 60, No.1, pp.125-136.

Seed, H. B. and Sun, J.I. (1989) "Implications of Site Effects in the Mexico City Earthquake of September 19, 1985 for Earthquake Resistant Design Criteria in the San Francisco Bay Area of California", Earthquake Engineering Research Center Report No. EERC 89-03, University of California, Berkeley, March.

Seed, R. B., and Bonaparte, R. (1992) "Seismic Analysis and Design of Lined Waste Fills: Current Practice," *Proceedings, Stability of Slopes and Embankments - II*, Vol. 2, Special Geotechnical Publication No. 31, ASCE, pp. 1521-1545.

Sharma, H.D. and Goyal, H.K., (1991) "Performance of a Hazardous Waste and Sanitary Landfill Subjected to Loma Prieta Earthquake", Proceedings, 2nd Conference on Recent Advancements in Geotechnical Earthquake Engineering and Soil Dynamics, University of Missouri, Rolla, MO.

Singh, S. and Murphy, B. J., (1990), "Evaluation of the Stability of Sanitary Landfills," In: *Geotechnics of Waste Fills - Theory and Practice*, ASTM STP 1070, pp. 240-258.

Singh, S. and Sun, J. I. (1995), "Seismic Evaluation of Municipal Solid Waste Landfills," Proc. *Geoenvironment 2000*, ASCE Specialty Conference, New Orleans, Louisiana, 22-24 February.

Stewart, J.P., Bray, J.D., Seed, R.B., Sitar, N., (1994) "Preliminary Report on the Preliminary Report on the Principal Geotechnical Aspects of the January 17, 1994 Northridge Earthquake", Earthquake Engineering Research Center Report No. EERC 94-08, University of California, Berkeley.

Sun, J.I., Golesorkhi, R., and Seed, H.B., (1988) "Dynamic Moduli and Damping Ratios for Cohesive Soils", Earthquake Engineering Research Center Report No. EERC 88-15, University of California, Berkeley, August.

Vesic, A. S., (1970) "Test on Instrumental Piles, Ogeechee River Site," *Journal of Soil Mechanics and Foundations Division, ASCE, 96*, No. SM2, March.

Vesic, A. S., (1967) "Ultimate Loads and Settlement of Deep Foundations in Sand," *Proceedings, Bearing Capacity and Settlement of Foundations Symposium*, Duke University, Durham, N.C.

Seismic Stability Analysis for Geosynthetic Clay Liner Landfill Cover Placement on Steep Slopes

Larry R. Taylor[1], T. Max Pan[2], A.S. Dellinger[3] and Bing C. Yen[4]

Abstract

A final cover system incorporating a geosynthetic clay liner (GCL) was proposed for closure of the Toyon Canyon Landfill, a municipal solid waste landfill located in Los Angeles, California. The site's steep side slopes and location in a seismically active area, as well as the relatively low interface shear strength of GCLs, posed unique challenges to the design and analysis of the proposed cover system. This paper discusses the seismic stability analysis performed for the final cover design. Assessment of shallow sliding stability of the proposed GCL cover system for both static and dynamic conditions, evaluation of interface shear strength parameters and development of GCL performance criteria are presented. Key elements of the analytical approach include laboratory testing to evaluate GCL interface shear parameters at low normal stresses (<10 kPa) and utilization of accepted finite element models to estimate the seismic response of the refuse fill and cover for design earthquake events. Yield displacement (i.e., permanent down-slope deformation) at the soil/GCL interface is calculated using a Newmark-type analysis for a range of yield acceleration values derived from laboratory interface shear testing. By comparing calculated yield accelerations for various cover alternatives to the results of the deformation analysis, the expected performance of various cover alternatives is quantified and compared.

Introduction

State and federal regulations governing the closure and post-closure maintenance requirements for municipal solid waste landfills require placement of a final cover system to prevent infiltration of rainfall and surface water into the fill and thereby minimize leachate generation and to control landfill gas emissions. California prescriptive standards require a minimum 61 cm-thick compacted foundation layer overlain by at least 30 cm of clean soil

---

[1] Manager of Geotechnical Services, Bing Yen & Associates, Inc., 17701 Mitchell North, Irvine, CA 92714

[2] Project Manager, IT Corporation, 2355 Main Street, Irvine, CA 92714

[3] Project Engineer, City of Los Angeles, Bureau of Sanitation, Solid Waste Management Division, 419 S. Spring Street, Suite 800, Los Angeles, CA 90013

[4] President, Bing Yen & Associates, Inc., 17701 Mitchell North, Irvine, CA 92714

having a maximum hydraulic conductivity of 1x10⁻⁶ cm/sec and a surficial layer of clean vegetative cover at least 30 cm in thickness. Engineered alternatives to the prescriptive standard are permitted if the prescriptive standard is not feasible and the alternative is consistent with various performance goals. Additionally, both state and federal regulations require that the integrity of the cover and final face be maintained for both static and seismic conditions.

Toyon Canyon Landfill is a municipal solid waste (MSW) landfill operated by the City of Los Angeles. The landfill occupies about 364,000 m² (90 acres) of canyon lands within the boundaries of Griffith Park, the largest municipal park in southern California. The site is surrounded by park facilities including hiking and equestrian trails, picnic areas, a golf course, playgrounds and the Los Angeles Zoo. Approximately 15 million cubic meters of Class III waste was disposed in this deep canyon landfill between 1957 and 1985. Waste disposal at the site resulted in the current configuration consisting of a relatively flat top surface encompassing about 200,000 m² (50 acres) and a 120 meter high side slope encompassing about 164,000 m² (40 acres). The east-facing side slope consists of 10 benched sections with typical face slopes of 2:1 (horizontal:vertical). A photograph depicting the site shortly before closure is shown in Figure 1.

FIGURE 1. Aerial photograph of Toyon Canyon Landfill prior to interim closure. (Photo courtesy of Los Angeles County Department of Public Works)

In 1992, the City of Los Angeles began engineering design for the final closure of the landfill. Due to the unavailability of a local source of low permeability soil suitable for the prescriptive cover system, the design team began evaluating several alternatives including utilization of a flexible membrane liner (FML) and/or geosynthetic clay liner (GCL) in place of the low permeability soil layer. Further evaluation of anticipated postclosure settlement and overall constructability indicated that a GCL was the preferred alternative. However, designers were faced with several design challenges and unknowns related to use of available GCLs. GCLs, while commonly used in liner systems, have not been applied as cover materials on such steep slopes. The majority of available soil/GCL interface shear strength data are from shear tests performed to simulate much higher overburden stresses associated with liner applications and might not be applicable to the low normal stresses associated with a landfill cover. The GCL strength behavior and other site-specific design parameters needed to be fully studied and quantified before the feasibility of the proposed GCL cover could be adequately assessed.

Conceptual GCL Cover System

Several alternative cover systems incorporating a GCL membrane were preliminarily assessed. The final conceptual cover system, illustrated in Figure 2, includes a 60 to 120 cm thick compacted foundation layer comprised of in-place temporary cover soil, a geocomposite drainage blanket, an upper 15 to 30 cm of compacted foundation soil, a GCL membrane, a 30 cm thick layer of protective cover soil/vegetative cover, and a pre-seeded erosion control blanket.

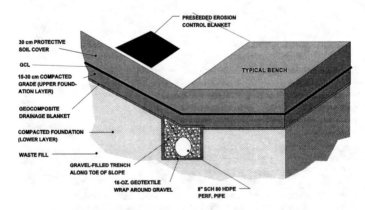

FIGURE 2 - Cross-section of conceptual final GCL cover system.

Assumptions for Limit Equilibrium Analysis of Sliding Stability

Initial analysis of shallow sliding of the final cover was based on a simple limit equilibrium method of analysis similar to that described by Chang, et.al.(1984). Factors of safety against shallow sliding and pseudo-static yield acceleration values were calculated for the conceptual cover system based on the following assumptions: 1) individual slope segments are of infinite length (i.e., no resisting force provided at toe of slope due to buttressing effect); 2) the soil/GCL interface is perfectly planar and parallel to slope face; 3) the cover soil is saturated and hydrostatic stress (buoyancy) may develop at the interface, and; 4) internal and interface shear strength parameters reported by GCL manufacturers are applicable at low normal stress levels.

The first three assumptions were considered to be realistically conservative. The typical slope length between benches is approximately 27 meters and shear resistance at the toe of individual slopes is negligible. In regard to planar slip surfaces, it is expected that actual construction operations and the landscape elements will create interfaces which contain asperities (e.g., horizontal and vertical irregularities and undulations) that result in a locally non-planar surface. In regard to hydrostatic stresses, it was assumed that the soil-GCL interface could become saturated due to infiltration of water from rain or irrigation or due to seepage of leachate from the waste prism, and hydrostatic stress could potentially act at the interface at some distance downslope along the slope face. A GCL with a nonwoven layer should provide reasonably good drainage at the interface due to the characteristics of the non-woven geotextile[1]. A FML is practically impervious and undrained shear conditions were assumed for this type of interface. Woven geotextile layers would not be expected to provide much drainage at the interface and undrained conditions were conservatively selected for these interfaces.

Internal shear strengths reported for the GCLs are typically greater than interface shear strengths and therefore the contact between the outer layer of the GCL and the soil presented the critical shear surface[2]. Data from manufacturers and in published literature indicated that interface shear tests are usually performed using normal stresses ranging from about 10 kPa to more than 1000 kPa. Adhesion values in the range of 2 to 10 kPa and interface friction angles ranging from 10 to 25 degrees have been reported for all of the products being considered. Based on the above general assumptions, static factors of safety ranging from about 1.2 to 3.2 were preliminarily calculated for critical interfaces of various candidate GCLs using post-peak (residual) interface shear strength parameters based on the manufacturers' data. Yield accelerations corresponding to the above static safety factors ranged from about 0.2 to 1.6 g.

Use of manufacturers reported interface shear strength parameters might provide either conservative or non-conservative results depending on the applicability of the strength parameters to low normal stresses (i.e., less than about 10 kPa). It is important to recognize that for the low normal stress conditions of the final cover, the calculated factor of safety and yield acceleration are extremely sensitive to adhesion values at the interface. In light of the

---

1 Typical hydraulic conductivity values for non-woven geotextiles are in the range of 5×10^{-1} to 10^{-2} cm/sec, which is several orders of magnitude greater than that of the final soil cover. Given this condition, the non-woven geotextile would be expected to provide sufficient drainage to prevent the development of hydrostatic stresses at the interface.

2 For stitch-bonded and needle-punched GCLs only. The internal shear strength of an "unreinforced" GCL would be equal to that of the hydrated bentonite and would be a critical shear surface.

relatively low static factors of safety calculated for the various alternatives and the uncertainty associated with some of the higher safety factors, the design team determined that it was necessary to verify the applicability of manufacturers' interface shear strength parameters to low normal stress conditions.

GCL Interface Shear Strength at Low Normal Stress

An interface shear testing system was constructed specifically for this series of tests. The apparatus has five shear devices aligned on a 1.2 m x 2.4 m platform. The testing system was designed to allow simultaneous performance of long-term (creep) interface shear tests and minimize the total time required to acquire data at different shear stress levels. One of the shear devices was configured for either long-term tests at constant shear stress or short-term tests at constant displacement rate. Each shear device was constructed in general compliance with ASTM Standard Test Method D 5321-92. The basic configuration of the shear devices consists of a stationary lower platform to which the GCL was attached and moving upper shear box containing soil. Normal stress was applied by attaching a 20 cm deep collar to the top of the shear box and filling it with sand.

The series of strain-controlled tests performed for this study involved one candidate GCL and soil from the Toyon Canyon Landfill site. The GCL consists of a layer of sodium bentonite sandwiched between two layers of non-woven polypropylene geotextile. The geotextile layers are needle-punched (geotextile fibers from one layer are pushed through the bentonite into the other geotextile layer) to increase internal shear strength.

Soil for the tests was composited from interim cover soil obtained by random sampling from the upper 50 cm at the Toyon site. The soil was screened through a No. 4 (4.75 mm) sieve and then blended using a large industrial mixer. The composited soil is a sandy lean clay (CL) with about 55% passing the No. 200 (75 μm) sieve. The liquid limit and plasticity index are 35% and 17, respectively. The maximum dry unit weight and optimum moisture content are 2.01 g/cm^3 and 11%, respectively. Direct shear tests on samples remolded in the lab to 90% relative compaction indicated a cohesion (c) of 13 to 15 kPa and friction angle (ϕ) of 31°.

Test Procedures

A specimen of GCL was cut from a factory-supplied roll and glued to the fixed lower platform using a two-part epoxy. Care was taken to avoid creating wrinkles in the GCL. Soil was compacted into the shear box at a dry unit weight of about 1.81 g/cm^3 and a moisture content of about 14% (90% relative compaction). Compaction was performed by placing the soil into the shear box in two lifts and compacting with 100 blows per layer using a 2.5 kg rammer with a 30 cm drop. The compacted soil was trimmed to be flush with the top of the shear box. PVC sheeting was placed on top of the soil, the extension collar was placed on the shear box, and dry sand was placed in the collar to apply the desired normal stress at the soil-GCL interface. The test platform was flooded with tap water to soak the interface for a minimum period of 24 hours prior to testing. The water level was maintained at a depth of about 1 cm above the soil-GCL interface throughout the test.

Short-term tests were performed in general conformance with the methods described in ASTM D 5321-92. Shear force was applied at a constant rate of displacement using a set of hydraulic cylinders. A controlled displacement rate of 1 mm/min was achieved using a flow control valve attached to both hydraulic cylinders in parallel. Shear force was measured using a strain-gauged load cell. Shear displacement was measured with a linear variable displacement transformer (LVDT) and with an analog dial gauge attached to the back of the shear box. Load cell and LVDT readings were digitally recorded on a PC-based computer data acquisition system every 15 seconds during the test (corresponding to readings about every 0.25 mm of displacement). Shearing was continued until a shear displacement of at least 4 cm was reached. Following each test, the test device was disassembled and the shear surface and GCL were visually examined.

Sample preparation for the long-term tests was identical to that for the short-term tests. A normal stress of 6 kPa was applied to each specimen. After the test specimens were soaked for 24 hours, an initial displacement reading was taken from a dial gauge attached to the back of the shear box. A dead weight was then placed on a cable assembly attached to the shear box to apply a constant shear stress to the test specimen. Constant shear stresses ranging from about 30% to 100% of the average short-term residual interface shear strength were used. Data for the long-term tests were collected for at least 1000 hours elapsed time and until the rate of displacement achieved a steady-state value.

Results of Interface Shear Tests

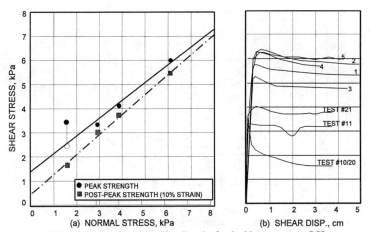

FIGURE 3 - Interface Shear Test Results for double non-woven GCL

Results of the short-term interface shear tests are presented in Figure 3. Five tests (identified as #1 through #5 in Fig. 3b) were performed at the design normal stress of 6.14 kPa. Test # 10 was performed on a specimen which had been subjected to more than 1800 hours of creep at 4.0 kPa constant shear stress. The normal load was decreased to 1.6 kPa and the sample was sheared at a constant rate of about 1 mm/minute. After shearing the specimen to about 10% strain, the normal stress was increased to 2.9 kPa and the sample was sheared to anadditional 10% shear strain (Test #11). This same sequence was followed for a test specimen which had been subjected to creep loading at 2.3 kPa, using normal stress increments of 1.6and 3.8 kPa (Tests #20 and 21). The displacement versus shear stress curves for all of the tests indicate that peak strengths are achieved at small displacements and with the exception of the 1.7 kPa tests, shearing resistance does not drop off appreciably within the first 4-5 cm of displacement and the shape of the curves suggests elasto-plastic behavior with small yield displacement. The 1.7 kPa tests also had small peak displacements but had significant strength reduction within the initial 2 cm of displacement.

Post-peak test results are summarized and compared to manufactures' reported values in Figure 4. The data indicate that the GCL has a nonlinear failure envelope and that more than one shearing mechanism governs the interface strength. At low normal stress, the post-peak failure envelope shows a small adhesion (approx. 0.3 kPa) and a friction angle of about 40°. The friction angle flattens to about 15° at normal stresses greater than 30 kPa, which equates

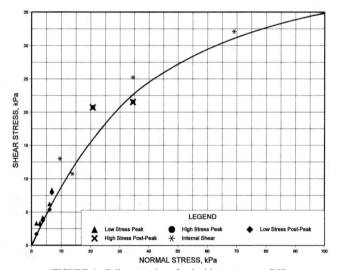

FIGURE 4 - Failure envelope for double non-woven GCL

to a soil depth of roughly 1.5 meters. Also plotted on Figure 4 are results of internal shear (i.e., the strength provided by the bentonite and reinforcing fibers) tests performed by others on the same GCL. The tests performed for this study indicate that the available shear strength for the design slope condition is 5.8 kPa. This is nearly 30% lower than the 8.0 kPa shear strength predicted using the manufacturer's reported strength parameters (ϕ=22° and c_a=5.5 kPa). Use of the low stress strength parameters reduced the calculated static factor of safety from 3.2 to 1.8 and reduced yield acceleration from 1.6g to 0.47 g for this interface condition. While low stress testing of other interfaces was not included in this study, it is reasonable to expect that similar behavior may exist for other products.

Long-Term Creep Test Results

Results of the stress-controlled tests are not directly relevant to the seismic stability analysis, but are presented here for general interest. Test data for the long-term interface shear tests are presented in Figure 5. Displacement rates for all samples were generally stable following 100 to 200 hours and reached steady-state conditions following about 500 to 800 hours.

Steady-state creep rates for the range of shear stresses applied (1.7 to 6 kPa) ranged from about 10^{-4} to 10^{-3} mm/hour. There appears to be a reasonable good correlation between shear stress and the logarithm of creep rate. For the typical slope angle (2 horizontal:1 vertical) and the proposed 30 cm vegetative cover configuration proposed for Toyon Canyon Landfill, the static shear stress at the soil-GCL interface will be approximately 2.8 kPa.

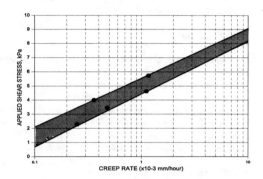

FIGURE 5 - Stress-controlled creep test results for double non-woven GCL.

Based on the results of the creep tests on the double non-woven GCL, it is estimated that steady-state downslope creep due to static shear stress will be on the order of 3×10^{-4} mm/hour, or about 3 mm/year.

Seismic Response and Stability Analysis

Toyon Canyon Landfill is situated in an area of high seismic activity. A thorough presentation of site seismicity is outside the scope of this paper and only the design earthquake events are discussed in the following paragraphs. During the landfill's extended lifetime, peak ground accelerations at the site are likely to exceed the yield accelerations calculated by limit equilibrium methods. Consequently, factors of safety against shallow sliding calculated by pseudostatic limit equilibrium methods are reduced to less than unity under seismic loading conditions for most or all of the GCL cover alternatives. However, pseudostatic analyses assume that a mass will "fail" if an earthquake-induced horizontal acceleration exceeds the calculated yield acceleration. In reality, a slope may be subjected to repeated pulses of acceleration and stress that may temporarily exceed the static failure strength along a failure surface. Each time the yield strength is exceeded, the mass defined by the failure surface will experience plastic deformation. The total amount of deformation experienced by the mass depends on the number and frequency of transient pulses which produce stresses greater than the yield strength of the failure surface.

In order to more realistically estimate the nature and magnitude of shallow sliding that may occur as a result of earthquake loading conditions, the design team undertook an analysis of seismically-induced permanent deformation along the interface of the final cover soil and the

GCL based on methods developed by Newmark (1965), Franklin & Chan (1977), Makdisi & Seed (1978) and others. Our analysis used accepted two-dimensional finite element models to predict seismic response at the face slope (soil/GCL interface). Yield displacement (i.e., permanent down-slope deformation) at the fill/GCL interface was then calculated using a Newmark-type analysis for a range of yield acceleration values. By comparing calculated yield accelerations for various cover alternatives - derived from laboratory interface shear testing - to the results of the deformation analysis, the expected performance of various cover alternatives could be quantified and compared.

<u>Design Earthquakes</u>

Two earthquake time histories were used as input to the seismic response analysis to represent mid-field and far-field events. The seismic response analysis was based on estimated Maximum Credible Earthquakes (MCE) corresponding to the mid-field and far-field events and not on Maximum Probable Earthquakes (MPE). While a probabilistic seismic analysis has not been performed for this study, it is reasonable to assume that during the design life of the site several episodes of ground motion similar to that assumed for the mid-field event may occur. The probability of the site being subjected to more than one episode of ground motions as strong as those modeled by the assumed far-field event is thought to be low.

Mid-Field Earthquake

Mid-field earthquakes for two nearby fault zones were considered. One was the MCE for the Newport-Inglewood Fault Zone located 9 km south-southwest of the Toyon site and the other was the MCE for the San Gabriel Fault located 11 km north of the site. The California Division of Mines and Geology (CDMG) estimates the MCEs for these two faults to be magnitude (M) 7 and 7½, respectively. The potential ground acceleration level at the Toyon site resulting from either mid-field MCE was estimated using the attenuation relationship recommended by Sadigh, et al. (1986), which generally provides near upper-bound accelerations. The potential ground acceleration level at the Toyon site based on Sadigh's attenuation relationship, are estimated to be approximately 0.4g for either event. The events were modeled using an accelerogram recorded at the Santa Cruz (U.C. Santa Cruz/Lick Lab) strong-motion station during the October 17, 1989 Loma Prieta Earthquake (M=7.0) from a limestone bedrock site located 17km from the epicenter. The Santa Cruz record, which has a recorded peak acceleration of 0.44g and a duration of strong motion of about 15 seconds, was used without scaling.

Consideration was given to utilizing the strong motion record from the nearby Griffith Park Observatory for the January 17, 1994 Northridge Earthquake (M=6.7) as a model mid-field event for the site. However, processed (digitized and corrected) data from the Griffith Park Observatory free-field station was not available at the time the analyses were performed. Based on the reported free-field peak accelerations from nearby strong motion stations and review of a preliminary accelerogram from the Griffith Park Observatory station, the assumed peak acceleration of 0.44g appeared to be reasonable and seismic response of the Toyon site calculated using either the Loma Prieta record or the Griffith Park record of the Northridge quake were expected to be similar.

The Northridge quake provided some interesting observations relevant to this seismic response analysis. The Toyon site is located approximately 24 km southeast of the epicenter of the quake, with a hypocentral distance of about 28 km from the main shock. Based on observations and the on-site engineers report, no visible significant damage to the face slope at Toyon could be attributed to the Northridge event.

The Northridge quake was recorded at many strong motion stations in the general and immediate vicinity of Toyon Canyon, including several stations located within a few kilometers of the site. It is well recognized that the Northridge Earthquake produced stronger near-field and mid-field response than would be predicted by most recent attenuation relationships, such as that developed by Boore, et al. (1993) based on a worldwide earthquake database. At an epicentral distance of 24km, a horizontal acceleration of about 0.15g would be predicted using Boore's attenuation relationship. A peak horizontal acceleration of 0.29g was recorded during the Northridge Earthquake at a USGS free-field strong motion recording station located at Griffith Park Observatory about 3.7 km south of the landfill. A CDMG station located 4 km north of the site at the base of a 6-story building constructed on alluvium recorded a peak horizontal acceleration of 0.35g. The nearest free-field instrument at a rock site, located at Pacoima-Kagel Canyon approximately 17 km northeast of the Northridge epicenter, recorded a peak horizontal acceleration of 0.44g. The largest free-field peak acceleration, recorded at the 7-story Los Angeles Universal Hospital located 36km from the epicenter, was 0.49g. Although the large lateral ground motions associated with the Northridge Earthquake are considered to be somewhat anomalous by some researchers, the potential for ground motions significantly larger than those predicted by many accepted attenuation relationships cannot be discounted. Based in part on this information, we elected to use attenuation relationships which provide near upper bound peak accelerations for the Toyon seismic analyses.

Far-Field Event

The San Andreas Fault, located approximately 35 km northeast of the site, was selected as the far-field event. CDMG estimates the Maximum Credible Earthquake (MCE) on the San Andreas Fault will generate a M=8½ event. Based on Sadigh's attenuation relationship, a peak horizontal bedrock acceleration of 0.27g at the Toyon site was calculated for this event. An artificial accelerogram simulating a free-field strong motion record (Caltech Artificial Earthquake - A1) was selected to represent the far-field event. The accelerogram models a relatively energetic low-frequency, long duration (35 seconds of strong motion) event. Peak acceleration was scaled to 0.27g using Sadigh's attenuation relationship.

Seismic Response Analysis

FEADAM (Duncan, et al., 1980) was used to estimate the distribution of stresses, strains and displacements in the refuse fill. FEADAM uses a nonlinear incremental finite element method of analysis and can utilize either linear elastic or stress dependent, non-linear elastic material properties. For this analysis, the refuse fill was modeled as a stress-dependent nonlinear elastic material. Due to the lack of reliable site-specific properties for the Toyon refuse fill, material properties for the analysis were assumed on the basis of published information on measured refuse strength properties (Singh & Murphy, 1990, Siegel, et.al., 1990, Landva, et.al., 1984). Finite element meshes were generated for three cross-sections representing the maximum, intermediate and moderate fill thickness. A layered construction sequence was used in the FEADAM analysis, with refuse fill placement simulated by sequentially "turning on" six layers of elements. The cover soil cover was modeled as a separate layer with different material properties.

QUAD-4 (Idriss, et al., 1973) was utilized to evaluate the seismic response of the landfill to input ground motions representing the far-field design event (San Andreas Fault) and the mid-field design event (San Gabriel Fault or Newport-Ingelwood Fault). QUAD-4 is a two-dimensional finite element program for evaluating the seismic response of soil structures using equivalent linear strain-dependent modulus and damping properties. QUAD-4 calculates time histories of acceleration for every element in the cross-sectional mesh.

<u>Refuse Fill and Bedrock Properties</u>

The refuse fill was assumed to have a bulk unit weight of 1.28 k/cm³, a friction angle of 35° and a cohesion of 16.8 kPa. The assumed strength parameters correspond to the upper mid-range of refuse strength parameters reported in the literature. The bedrock underlying the refuse was assumed to have a shear wave velocity (V_S) of 600 meters/second and G_{max}=830 MPa. The assumed dynamic properties of the refuse fill are plotted in Figure 6. Shear modulus (G_{max}) and strength (S_u) vary with mean confining stress, σ_0 (calculated using FEADAM). Strain-dependent damping and modulus values were assumed to follow the relationships suggested by Singh & Murphy (1990), i.e., the statistical average of curves for peat and clay.

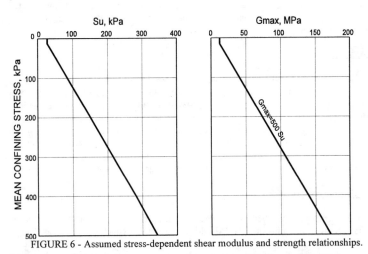

FIGURE 6 - Assumed stress-dependent shear modulus and strength relationships.

The authors note that one of the greatest areas of uncertainty in this type of analysis relates to selection of appropriate strength, modulus and damping values for the relatively deep refuse fill (approximately 100 meters at Toyon). The amount of detailed data regarding the static and dynamic strength and behavior of refuse fill is limited and available data indicate substantial variability. The age of the fill, amount of soil used for daily cover, types of waste accepted by the landfill, moisture conditions and other factors may greatly influence shear strength characteristics. Readers are cautioned that selection of representative parameters should be based on appropriate assumptions and that a sensitivity analysis should be performed to assess the influence of various assumptions on the results of the analyses. For these analyses, the design team assumed relatively dense, high strength/high modulus refuse fill because it provided more conservative results (e.g., greater response at the soil/GCL interface) than did an assumed low modulus and lightweight fill. While these assumptions may lead to overestimating seismic response at the surface, this is prudent when data to the contrary do not exist.

Comparison of calculated response histories to the input accelerograms indicated that the ground motions may be either attenuated or amplified, depending on location, depth of the landfill waste prism, predominant frequency, and other variables. Generally speaking, peak acceleration response at the landfill surface was amplified from peak accelerations in the bedrock for the far-field event, with 0.27g peak bedrock accelerations producing peak accelerations as high as 0.5g at the landfill surface. Peak surface accelerations in response to the mid-field event were generally equal to or less than the assumed bedrock acceleration of 0.44g. The later result is consistent with recent measurements made at the Operating Industries, Inc. (OII) Landfill in Monterey Park where peak accelerations of approximately 0.248g and 0.255g were recorded during the Northridge Earthquake at the base (bedrock) and top of the landfill, respectively.

Seismic Deformation Analysis

The seismic response analysis provided acceleration time histories at nodal points along the surface of the slope. Double-integration of the acceleration histories at nodal points along the slope face was performed to calculate permanent displacements for assumed yield accelerations of 0.0g, 0.1g, 0.2g and 0.3g and 0.5g. General results of the analysis in terms of the maximum displacements calculated for several generic soil/GCL interfaces are summarized in Table 1.

TABLE 1 - Summary of Sliding Stability and Deformation Analysis Results

Generic Interface	Strength Parameters		Static Factor of Safety	Yield Accel. (g)	Max. Disp. for Far-Field MCE (cm)	Ref.
	ϕ_{pp} (deg)	c_a (kPa)				
Soil/Nonwoven Geotextile	36	0.3	1.58	0.47	<1	BYA, 1994
Soil/Nonwoven Geotextile	22	5.5	3.2	1.6	0	Mfg. test data
Soil/Woven Geotextile	25	2.4	1.44	0.36	15	Mfg. test data
Soil/Roughened HDPE	29	10.5	4.80	3.04	0	Mfg. test data
Soil/Bentonite	10	2.6	1.24	0.19	55	Mfg. test data

In order to estimate the nature of deformation, displacements were calculated at nodal points spaced approximately 30 meters apart (horizontally) along each of the three FEM cross-sections and plotted to interpolate contours of equal down-slope deformation on the face slope. In developing the contour maps, deformation was assumed to be generally symmetric across the maximum cross-section A-A'. Figure 7 shows an example of deformation contours corresponding to the assumed far-field event. By comparing the calculated minimum yield acceleration corresponding to a given interface to the parametric deformation analysis results, seismic-induced slope displacements were estimated for each cover alternative.

The analyses indicate that maximum displacements are likely to occur near the abutments on the upper portions of the slope and in the central portion of the uppermost benches. The somewhat unusual lobe shapes for the 0.2g contours on Figure 4 are primarily related to variations on refuse fill thickness and depth to basement rock in the former natural canyon. Deformation contours are generally consistent with and proportional to refuse fill thickness.

The most likely manifestation of deformation will be formation of tension cracks at or near the top of individual benched slopes, requiring cosmetic repair. The uppermost soil-GCL interface represents the weakest layer and therefore the displacements should occur between the cover soil and GCL. The integrity of the GCL should not be compromised by the potential displacement of cover soil. Seismically-induced slippage of the cover soil may also cause localized damage to components of the gas collection, irrigation and surface drainage systems.

Conclusions

Available methods for estimating the seismic response of earth structures were utilized together with a simple limit equilibrium method to analyze the sliding stability and the magnitude and distribution of seismic-induced deformation for a GCL landfill cover system on steep slopes. This approach offered a relatively straightforward method of assessing a cost-effective alternative to the regulatory-prescribed cover for closure of a landfill with an existing 2:1 face slope for which the geometry could not be altered.

Due to the lack of available data regarding the interface shear strength parameters for normal stresses within the range of interest for this study, a number of low stress interface shear tests were performed. These tests indicate that, for the GCL tested, available shear strength at normal stresses below 10 kPa is significantly lower than that predicted by the manufacturers data. The results emphasize the importance of selecting the GCL cover component based on the results of carefully specified conformance tests that include interface shear testing which addresses the anticipated field conditions with respect to the range of normal stress, saturation and drainage conditions, test specimen orientation and preparation, strain rate, and other factors.

The results of the analysis indicate that maximum permanent sliding deformation in response to the design earthquake may range from less than one centimeter to about 0.5 meter for available GCLs. By establishing limits on the maximum allowable seismically-induced deformation, performance criteria for candidate GCLs can then be established. For Toyon, a maximum allowable deformation of about 15 cm (6 inches) has been preliminarily established on the basis of estimated repair requirements. Therefore, candidate GCLs must be demonstrated to possess interface shear strength parameters which render a maximum

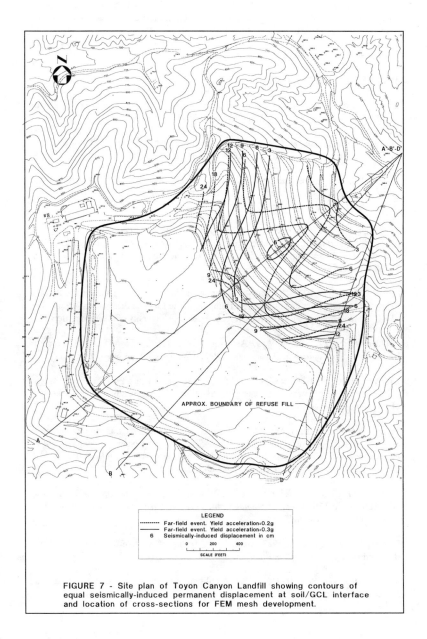

FIGURE 7 - Site plan of Toyon Canyon Landfill showing contours of equal seismically-induced permanent displacement at soil/GCL interface and location of cross-sections for FEM mesh development.

yield acceleration of approximately 0.35g for the assumed slope conditions. Specifications for conformance tests were developed to enable a reasonable and consistent basis for comparing candidate GCL products. In addition, it was possible to develop estimates of the cover maintenance requirements and costs to address post-closure maintenance issues mandated by landfill closure regulations.

APPENDIX - REFERENCES

Bing Yen & Associates, Inc., 1993, "Direct Shear Test Result of Existing Cover Soil, Toyon Landfill," prepared for IT Corporation, January, 1993.

Chang, C.J., Chen, W.F., and Yao, J.T., 1984, "Seismic Displacements on Slopes by Limits Analysis," *Journal of the Geotechnical Division*, ASCE, New York.

Duncan, J.M., Wong, K.S., and Ozawa, K., 1980, "FEADAM: A Computer Program for Finite Element Analysis of Dams," Report No. UCB/GT/80-02, Department of Civil Engineering, University of California, Berkeley, December, 1980.

Franklin, A.G. and Chang, F.K., 1977, "Earthquake Resistance of Earth and Rockfill Dams, Report No. 5, Permanent Displacements of Earth Embankments by Newmark Sliding Block Analysis," U.S. Army Engineering Waterways Exp. Stat., Soils and Pavements Laboratory, Misell Paper S-71-17.

Idriss, I.M., et. al., 1973, "QUAD-4: A Computer Program for Evaluation the Seismic Response of Soil Structures by Variable Damping Finite Element Procedures," Report No. EERC 73-16, Earthquake Engineering Response Center, University of California, Berkeley.

I.T. Corporation, 1988, "Solid Waste Assessment Test (SWAT), Toyon Landfill," prepared for the Bureau of Sanitation, Department of Public Works, City of Los Angeles, June.

I.T. Corporation, 1994, "Final Closure and Postclosure Maintenance Plan for Toyon Canyon Sanitary Landfill," prepared for the Bureau of Sanitation, Department of Public Works, City of Los Angeles, September.

Joyner, W.B. and Boore, D.M., 1988, "Measurements, Characterization, and Prediction of Strong Ground Motion," editor Von Thun, *Proceedings: Conference on Earthquake Engineering and Soil Dynamics II - Recent Advances in Ground Motion Evaluation,* ASCE Geotechnical Special Publication No. 20, pp 43-102.

Landva, A.O., Clark, J.I., Weisner, W.R. and Burwash, W.J., 1984, Geotechnical Engineering and Refuse Landfills", *Sixth National Conference on Waste Management in Canada*, Vancouver, British Columbia

Landva, A.O. and Clark, J.I., 1987, "Geotechnical Testing of Wastefill", Report submitted to the London Institution of Civil Engineers

Makdisi, F.J. and Seed, H.B., 1978, "Simplified Procedure for Estimating Dam and Embankment Earthquake-Induced Deformations," *Journal of the Geotechnical Division*,

ASCE, Vol. 104, No. GT7.

Morris, D.V. and Woods, C.E., 1990, "Settlement and Engineering Considerations in Landfill and Final Cover Design," *Geotechnics of Waste Fills, Theory and Practice*, ASTM STP1070, American Society for Testing and Materials, Philadelphia, 1990.

Newmark, N.M., 1965, "Effects of Earthquakes on Dams and Embankments", Fifth Rankine Lecture, *Geotechnique*, Vol. 15, No.2, January, pp. 139-160

Sadigh, K., Egan, J., and Young, R., (1986), "Specifications for Ground Motion for Seismic Design of Long Period Structures (abs)," *Earthquake Notes*, 57, 13.

Singh, S. and Murphy, B., 1990, "Evaluation of the Stability of Sanitary Landfills," *Geotechnics of Waste Fills, Theory and Practice*, ASTM STP1070, American Society for Testing and Materials, Philadelphia, 1990.

SLIP DISPLACEMENTS OF GEOSYNTHETIC SYSTEMS UNDER DYNAMIC EXCITATION

Yegian, M.K.[1], M.ASCE and Harb, J.N.[2], S.M.ASCE

ABSTRACT

The dynamic response of geosynthetic interfaces commonly used in Municipal Solid Waste Landfills (MSWL) are investigated using a shaking table facility. Geosynthetic interfaces placed horizontally and on inclined surfaces were tested to simulate bottom and cover liner systems. This paper primarily focuses on the research results related to the slip displacements induced under harmonic base excitation. Normalized plots are presented that can be used to estimate the maximum (peak-to-peak) slip along horizontally placed geosynthetic interfaces. In addition, plots of normalized permanent slip -per cycle of harmonic excitation- along geosynthetic interfaces placed on various slope inclinations are presented. Example calculations are included.

INTRODUCTION

One of the critical components of a MSWL is the geosynthetic system utilized as an impermeable barrier. Under earthquake excitation, it is imperative that the integrity of such a system, hence that of the landfill itself, is maintained. Therefore, understanding the dynamic response of geosynthetic interfaces is an important area for research. During the past few years, the authors have been investigating the behavior of geosynthetic interfaces under dynamic excitation

---

[1] Professor and Chairman, Dept. of Civil Engrg., Northeastern Univ., Boston, MA 02115

[2] Graduate Student, Dept. of Civil Engrg., Northeastern Univ., Boston, MA 02115

using a shaking table facility. In this research program, the acceleration transmitted through various types of geosynthetic interfaces, as well as the slip displacements (hereafter referred to slip) developed along the interfaces have been investigated. Initial test results and observations have been reported by Yegian and Lahlaf (1991) and Yegian et al. (1995).

This paper presents results of shaking table tests performed on typical geosynthetic interfaces used in current practice for landfill design. The paper focuses primarily on the slip recorded along the interfaces tested. Normalized relationships are presented to predict slip under harmonic excitations. This research continues to evaluate slip along geosynthetic interfaces under earthquake excitations. The results from these tests will be published in the near future.

SHAKING TABLE FACILITY

Figure 1 shows a schematic drawing of the shaking table setup used in our research. The shaking table is powered by a hydraulic actuator that controls the horizontal displacement of the table. A geomembrane is fixed on the table upon which rests a plexiglass box used to hold soil or any other geosynthetic. Lead weights are added to increase the normal stress at the interface. The accelerations of the table and the box are measured by accelerometers. The relative displacement between the table and the box, hence slip along the interface, is measured by a Linear Variable Displacement Transducer (LVDT) as shown in Figure 1.

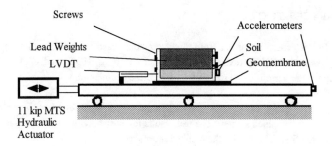

Figure 1.   Schematic Diagram of the Test Setup for Investigating the Dynamic Shear Properties of Geomembrane / Soil Interface.

To simulate landfill side slope conditions in the experiments, an adjustable-slope table is bolted on the main horizontal shaking table as shown in Figure 2. Test results reported in this paper correspond to both, horizontally placed as well as sloping interfaces.

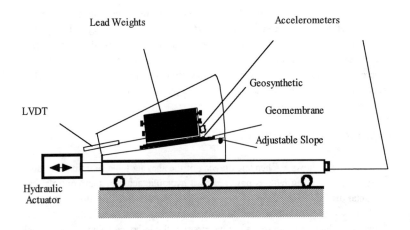

Figure 2.   Schematic Diagram of the Test Setup for Investigating the Dynamic Shear Properties of Geosynthetic Interfaces on a Slope.

INTERFACES TESTED

In this paper, test results from three commonly used geomembranes are presented, namely: smooth HDPE, textured HDPE referred to as HDT, and PVC.   Table 1 summarizes the various interfaces tested using these geomembranes.

As was stated earlier, the geosynthetic interfaces were tested horizontally as well as along inclined slopes. The test results for horizontally placed interfaces are presented first followed by the inclined interfaces.

Table 1.   Summary of the Various Geosynthetic Interfaces Tested.

Interface		Description
HDPE (Smooth)	Geotextile	60 MIL - HDPE (Gundline)/Polyfelt TS 700 (Nonwoven, continuous filament, needle punched geotextile)
	Geogrid	60 MIL - HDPE (Gundline)/Gundnet XL-14
	Gundseal (dry)	60 MIL - HDPE (Gundline)/Gundseal HD 30 k<E-12 cm/s, w=12%
	Soil-Bentonite Mixture (dry)	60 MIL - HDPE (Gundline)/Compacted Sand, 20% Silt and 5% Bentonite, K<E-7cm/s (next day after compaction)
	Soil-Bentonite Mixture (wet)	60 MIL-HDPE (Gundline)/Compacted Sand, 20% Silt and 5% Bentonite, K<E-7cm/s (right after compaction)
	Ottawa Sand	60 MIL-HDPE (Gundline)/Densified Ottawa Sand
HDT (Textured)	Geotextile	60 MIL-HDT (Gundline)/Polyfelt TS 700 (Nonwoven, continuous filament, needle punched geotextile)
	Soil-Bentonite Mixture (dry)	60 MIL-HDT (Gundline)/Compacted Sand, 20% Silt and 5% Bentonite, K<E-7cm/s (next day after compaction)
PVCs	Gundseal (dry)	30 MIL-PVC smooth (Palco)/Gundseal HD 30 k<E-12 cm/s, w=12%
	Geotextile	30 MIL-PVC smooth (Palco)/Polyfelt TS 700 (Nonwoven, continuous filament, needle punched geotextile)
PVCr	Gundseal (dry)	30 MIL-PVC rough (Fisher)/Gundseal HD 30 k<1e-12 cm/s, w=12%
	Geotextile	30 MIL-PVC rough (Fisher)/Polyfelt TS 700 (Nonwoven, continuous filament, needle punched geotextile)

RESULTS OF HORIZONTALLY PLACED GEOSYNTHETIC INTERFACES

Figure 3 shows typical acceleration time records of the table (base acceleration) and that transmitted through an HDT geomembrane to the compacted soil-bentonite mixture described in Table 1 (transmitted acceleration). From this figure, it is observed that there is a limiting acceleration that the HDT/Soil-Bentonite interface can transmit. Because of this difference between the base and transmitted accelerations, a relative displacement, slip, is induced along this horizontally placed interface. Figure 4 shows the slip measured during the test on HDT/Soil-Bentonite interface of Figure 3.

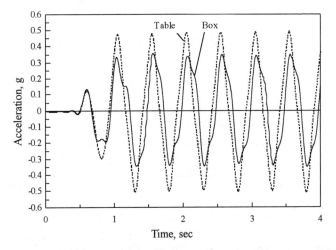

Figure 3.   Recorded Table (Base) and Box (Transmitted) Accelerations Versus
             Time, for HDT / Soil-Bentonite Mixture Interface on a Horizontal
             Surface.

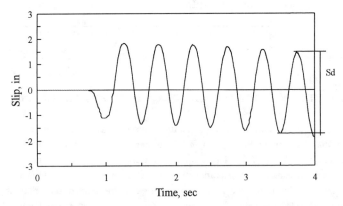

Figure 4.   Recorded Slip Versus Time and Maximum (Peak-to-Peak) Slip, S_d for
             HDT/Soil-Bentonite Mixture Interface on a Horizontal Surface

The results shown in Figure 4 indicate that, along a horizontally placed geosynthetic interface, under harmonic excitation, although the permanent slip at the end of a cycle is negligible, a maximum (peak-to-peak) slip S_d of about 3 inches has taken place during the excitation pulse. Thus, along a bottom liner of a landfill, maximum slip can be greater than permanent slip. Such peak-to-peak slip deformations along a bottom liner can be a concern, especially if the integrity of the leachate collection system is to be preserved.

It is clear that in a seismic response analysis of a landfill, the acceleration transmitted through the bottom liner, as well as the maximum slip along the liner are both of interest. In the following figures, the limited transmitted accelerations and maximum slips along typical geosynthetic interfaces are shown.

Figure 5 shows the amplitude of the transmitted acceleration versus the amplitude of the base acceleration for a number of interfaces that include smooth HDPE. The results in Figure 5 show that beyond a certain level of base acceleration, the transmitted acceleration is *smaller* than the base acceleration.

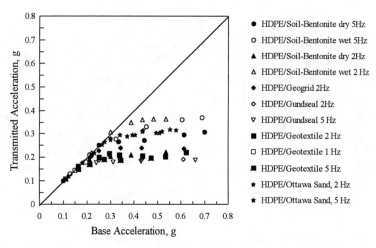

Figure 5.   Recorded Transmitted Acceleration Versus Base Acceleration for Smooth HDPE Interfaces.

We have examined the force-displacement relationships of these interfaces and have concluded that the interface characteristic is not perfectly rigid-plastic, as is commonly assumed, but rather elasto-plastic. Because of this, even under small levels of base acceleration, very small slips were recorded. Thus, it is difficult to define a yield acceleration from the acceleration records such as shown in Figure 5. Traditionally, yield acceleration is commonly defined as the maximum acceleration that can be transmitted through the geosynthetic interface. Alternatively, the yield acceleration is defined as that acceleration, if exceeded, slip deformations along the interface are induced.

Figure 6 shows the measured maximum slips along a smooth HDPE/geotextile interface for 1, 2, and 5 Hertz harmonic excitations. The results indicate that at small base accelerations, slip displacements are extremely small and difficult to be accurately recorded by the LVDT. At relatively large base accelerations, maximum slip increases approximately linearly with an increase in base acceleration. In this investigation, yield acceleration K_y is defined by extending the linear portion of the maximum slip curve to the left to intersect the horizontal axis that defines base acceleration. For the case of smooth HDPE/geotextile, this construction is shown in Figure 6. The results show that the yield acceleration, K_y - *the base acceleration beyond which measurable slip deformations are observed*- is about 0.16 g regardless of the frequency of the harmonic base motion.

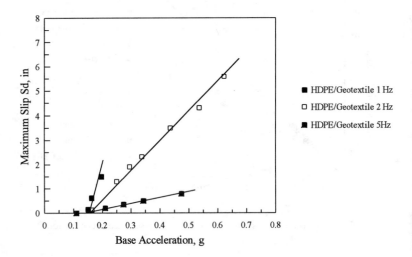

Figure 6.   Recorded Maximum Slip S_d Versus Base Acceleration for Smooth HDPE/Geotextile Interfaces.

Figure 7 shows, for purposes of comparison, the maximum slips, S_d measured for different smooth HDPE interfaces and at different frequencies of excitation.

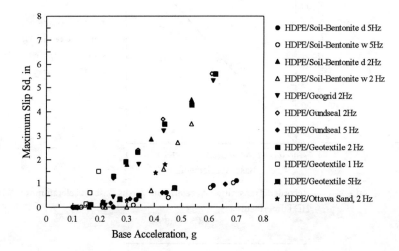

Figure 7.   Recorded Maximum Slip S_d Versus Base Acceleration for different
                 Smooth HDPE Interfaces.

Figures 8 and 9 show the transmitted accelerations and maximum slips for interfaces with textured HDPE (HDT) geomembrane.   Similar results for PVC geomembrane interfaces are shown in Figures 10 and 11.

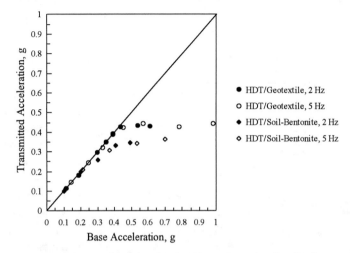

Figure 8.　　　Recorded Transmitted Acceleration Versus Base Acceleration for HDT Interfaces.

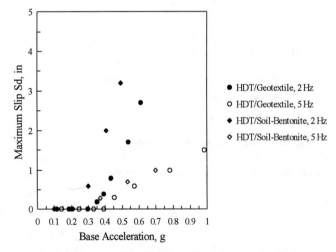

Figure 9.　　　Recorded Maximum Slip, S_d　Versus Base Acceleration for HDT Interfaces.

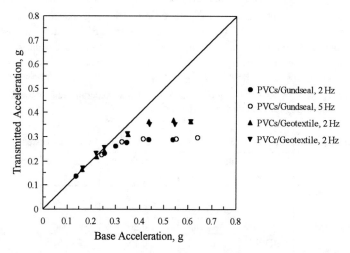

Figure 10.   Recorded Transmitted Acceleration Versus Base Acceleration for PVC Interfaces.

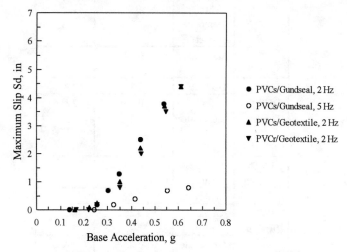

Figure 11.   Recorded Maximum Slip, S_d Versus Base Acceleration for PVC Interfaces.

Application in Practice

The maximum slips S_d measured along horizontally placed geosynthetic interfaces were normalized to provide relationships useful in seismic design practice for landfills. The test results indicate that the type of interface, the level of base acceleration and the frequency of the base motion have significant influence on the maximum slip induced. For the purpose of preparing normalized plots of maximum slip, the dynamic shear strength properties of the tested interfaces are represented by the yield acceleration K_y as defined previously in this section. Table 2 lists the experimentally determined yield accelerations K_y for the interfaces tested.

Table 2   Yield Acceleration, K_y for the Geosynthetic Interfaces Tested

Interface		K_y, Yield Acc. (g)	Frequency (Hz)
HDPE (smooth)	Geotextile	0.16	1, 2 & 5
	Geogrid	0.20	2
	Gundseal (dry)	0.15	2 & 5
	Soil-Bentonite (dry)	0.18	2 & 5
	Soil-Bentonite (wet)	0.3	2 & 5
	Ottawa Sand	0.27	2 & 5
HDT (Textured)	Geotextile	0.36	2 & 5
	Soil-Bentonite (dry)	0.26	2 & 5
PVCs (Smooth)	Gundseal (dry)	0.23	2
PVCr (Rough)	Geotextile	0.22	2
	Gundseal (dry)	0.26	2
	Geotextile	0.23	2 & 5

Yegian et al. (1991), in their research on earthquake-induced permanent deformations of earth dams, defined a normalized relative displacement function as:

$$D_n = \frac{D_r}{K_a . T^2 . N_{eq}} \tag{1}$$

where D_n is the non-dimensional normalized relative displacement parameter, D_r is the relative displacement, K_a is the acceleration of a rigid support, T is the period of the base motion and N_{eq} is the equivalent number of uniform pulses of the motion. Following this approach, the maximum slips, S_d along the tested geosynthetics were normalized as in Equation 2

$$S_n = \frac{S_d}{K_a . T^2} \tag{2}$$

where S_n is the normalized (non-dimensional) maximum slip, S_d is the actual measured maximum slip, K_a is the base acceleration, and T is the period of the base motion.

The maximum slips shown in Figures 6, 7, 9, and 11 were normalized using Equation 2 and then plotted in Figure 12, as a function of the ratio of K_y (from Table 2) and K_a. This plot shows a definite trend of the normalized slip, S_n decreasing with increasing K_y/K_a ratio. A regression analysis was performed on the data shown in Figure 12, which resulted in Equation 3.

$$S_n = \frac{S_d}{K_a . T^2} = 0.128 - 0.135\left(\frac{K_y}{K_a}\right) + 0.005\left(\frac{K_y}{K_a}\right)^2, \quad r^2 = 0.95 \tag{3}$$

Equation 3 can be used to estimate the mean value of maximum slip S_d along a geosynthetic interface. Example applications follow:

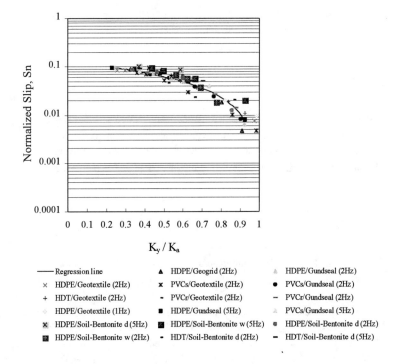

Figure 12.   Normalized Slip, S_n Versus K_y / K_a for the Different Types of
                Geosynthetics Tested.

Examples (a) and (b)

a) Consider:

      HDT/geotextile interface horizontally placed
      Base motion $K_a = 0.4g$
      Period of motion, T = 0.5 to 1.0 second

Therefore:

K_y = 0.36 g from Table 2

K_y / K_a = 0.36/0.4 = 0.9

S_n = 0.01055 from Figure 12 or Equation 2

$S_d = S_n . K_a . T^2 =$ 0.4 to 1.6 inches for T = 0.5 to 1.0 second

b) Consider:

Smooth HDPE/geotextile

Period of motion, T = 0.5 to 1.0 second

Base motion K_a = 0.4g

Therefore:

K_y = 0.16 g from Table 2

K_y / K_a = 0.16/0.4 = 0.4

S_n = 0.0748 from Figure 12 or Equation 2

$S_d = S_n . K_a . T^2 =$ 3.0 to 12.0 inches for T = 0.5 to 1.0 second

It is evident that maximum slip, S_d along bottom liners of landfills depends on the type of interface, the base acceleration, and, very importantly, the period of the motion. Typically, for textured HDT/geotextile slip will be on the order of a few inches. For smooth HDPE/geotextile, maximum slip can be greater than a foot depending on K_a and T. The implications of such slip on the integrity of the bottom liner and the leachate collection system is not yet well understood and await further research.

RESULTS ON INCLINED GEOSYNTHETIC INTERFACES

To evaluate the dynamic response of geosynthetic interfaces placed on side slopes of landfills, the experimental setup shown in Figure 2 was used. Figure 13 shows a typical acceleration record from a shaking table test on HDT /Soil-Bentonite interface placed on 4H:1V slope. Figure 14 shows the measured permanent slip along the interface as a function of time. It is evident from the record that the behavior of the inclined interface is more complex than a horizontal one. Depending on the interface and the level of base acceleration, slip occurred in both directions, down-slope, as well as up-slope.

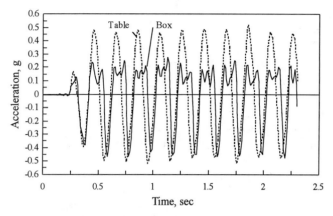

Figure 13.   Recorded Base and Transmitted Accelerations Versus Time for HDT /
              Soil-Bentonite Mixture Interface on an Inclined Surface (4H:1V).

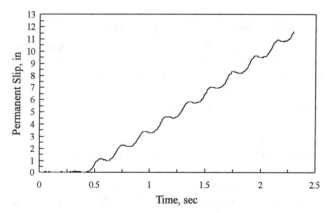

Figure 14.   Recorded Permanent Slip Versus Time for HDT / Soil-Bentonite Mixture
              Interface on an Inclined Surface (4H:1V).

Difficulties were encountered in defining a single parameter K_y to describe the
yield acceleration of the inclined interface. Newmark (1965) defined $K_{y,\theta}$ parallel
to the slope as function of the yield acceleration on a horizontal plane, K_{yh} , and
the slope inclination θ.

$$K_{y,\theta} = Cos\theta . K_{y,h} \mp Sin\theta \qquad\qquad (4)$$

Considering that in our experiments the base acceleration is horizontal and not parallel to the slope, Equation 4 can be modified by rewriting the equations of static equilibrium and including the inertial forces similar to a Newmark analysis. Equation 5 expresses $K_{y,\theta}$ for a horizontal base motion, as function of θ.

$$K_{y,\theta} = \frac{Cos\theta . K_{y,h} \mp Sin\theta}{Cos\theta \pm Sin\theta . K_{y,h}} \qquad\qquad (5)$$

It is noted that for practical slopes that are encountered in landfills (up to 3H:1V), the two equations yield very close values of $K_{y,\theta}$. However, when the $K_{y,\theta}$ values calculated from Equations 4 or 5 were compared with values estimated from the measured data (using similar procedure as that described in the previous section) , significant discrepancies were observed. The likely reasons for these discrepancies are:

1) It is very difficult to define $K_{y,\theta}$ from the data when there is both down-slope and up-slope slip. Also, as shown in Figure 13, the down-slope acceleration varies during the duration of the half-pulse.
2) The equations for $K_{y,\theta}$ are based on the assumption of a rigid-plastic interface shear behavior, and do not account for the dynamic behavior of the system associated with the more realistic elasto-plastic behavior of the interface.

Figure 15 shows a comparison between the measured and the calculated $K_{y,\theta}$ for selected interfaces. The data clearly indicate that it would be difficult to uniquely relate K_y for an inclined interface with that of a horizontal interface and the slope angle through the use of Equations 4 or 5. Since in current design practice for landfills, a limited number of geosynthethic interfaces are utilized along a few distinct slope angles, it was decided to present the test results in terms of normalized permanent slip as a function of typical slope angles, and for typical interfaces commonly used in landfills.

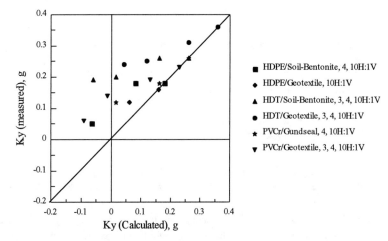

Figure 15.   Comparison Between the Measured and Calculated K_y for Different Interfaces and slope angles.

Figures 16 through 20 show the measured permanent slips, PS_d (inches/cycle) for variety of interfaces and slope inclinations.

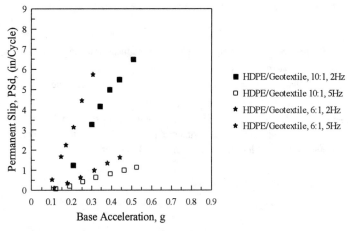

Figure 16.   Permanent Slip per Cycle, PS_d Versus Base Acceleration for HDPE/Geotextile Interface

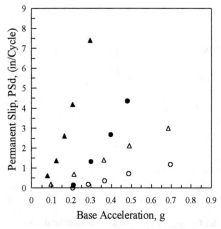

Figure 17.   Permanent Slip per Cycle, PS_d, Versus Base Acceleration for
HDPE/Soil-Bentonite dry Interface

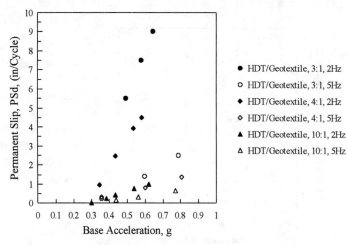

Figure 18.   Permanent Slip per Cycle, PS_d, Versus Base Acceleration for
HDT/Geotextile Interface

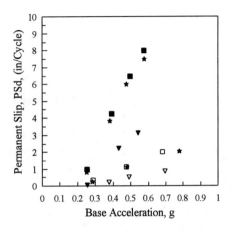

Figure 19.   Permanent Slip per Cycle, PS_d, Versus Base Acceleration for
              HDT/Soil-Bentonite Mixture dry Interface

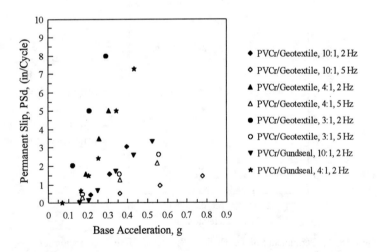

Figure 20.   Permanent Slip per Cycle, PS_d, Versus Base Acceleration for PVC
              Interfaces.

Application in Practice

To render the experimental data useful in engineering practice, the permanent slip, PS_d, data were normalized as was described in the preceding section.

$$PS_n = \frac{PS_d}{K_a T^2} \qquad (6)$$

where PS_n is the non-dimensional normalized permanent slip per cycle, PS_d is the permanent slip per cycle, K_a is the base acceleration, and T is the period of the motion.

The normalized permanent slips, PS_n were plotted versus the base acceleration K_a as shown in Figures 21 through 25. These figures can be used to estimate the permanent slip per cycle of excitation along typical geosynthetic interfaces placed in various slope inclinations.

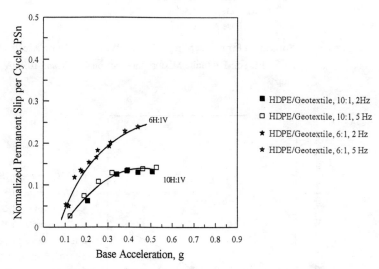

Figure 21.   Normalized Permanent Slip per Cycle, PS_n, Versus Base Acceleration for HDPE/Geotextile Interface on 6H:1V and 10H:1V Slopes.

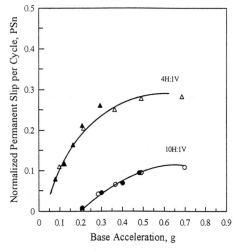

Figure 22.   Normalized Permanent Slip per Cycle, PS_n, Versus Base Acceleration
              for HDPE/Soil-Bentonite Mixture Interface on 4H:1V and 10H:1V
              Slopes.

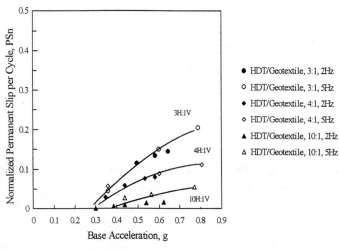

Figure 23.   Normalized Permanent Slip per Cycle, PS_n, Versus Base
              Acceleration for HDT/Geotextile Interface.

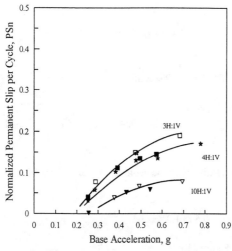

Figure 24.   Normalized Permanent Slip per Cycle, PS_n , Versus Base Acceleration
for HDT/Soil-Bentonite dry Interface.

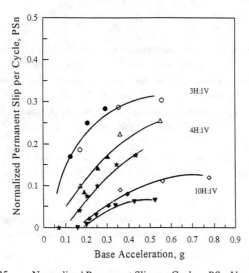

Figure 25.   Normalized Permanent Slip per Cycle, PS_n , Versus Base Acceleration
for PVC Interfaces.

Example applications follow:

a) Consider:

        HDT/geotextile interface when the geotextile is not anchored

        Slope 3H:1V

        Base motion K_a = 0.4g

        Period of motion, T = 0.5 to 1.0 second

Therefore:

        PS_n = 0.07 g from Figure 23.

        $PS_d = PS_n\, K_a T^2$ = 2.7 to 11 inches / cycle for T = 0.5 to 1.0 second

b) Consider:

        Smooth HDPE/geotextile when the geotextile is not anchored

        Slope 6H:1V

        Base motion K_a = 0.4g

        Period of motion, T = 0.5 to 1.0 second

Therefore:

        PS_n = 0.23 from Figure 21.

        $PS_d = PS_n\, K_a T^2$ = 9 to 36 inches / cycle for T = 0.5 to 1.0 second

SUMMARY

Shaking table tests were performed to investigate the relative displacements (slip) induced along geosynthetic interfaces under harmonic excitation. Based on the test results, normalized plots are presented that can be used to estimate the maximum slip along horizontally placed geosynthetic interfaces, as well as the permanent slip induced along inclined interfaces. Also, from the test results, the following observations and conclusions are made:

1) On horizontally placed geosynthetics, yield acceleration can best be defined as the acceleration beyond which measurable slip is induced along the interface.

2) For horizontal geosynthetic interfaces, maximum (peak-to-peak) slip, not permanent slip, is of concern. Hence, maximum slip along the bottom liner of

a landfill should be limited in design to ensure the integrity of the bottom leachate collection system.

3) Yield acceleration for inclined geosynthetic interfaces, calculated based on Newmark's equation are not in agreement with measured values from the shaking table tests. This discrepancy is attributed to elasto-plastic behavior of the interface as well as non-constant yield acceleration measured during slip in the down-slope direction. For this reason, it was deemed more appropriate to present normalized permanent slip as a function of base acceleration, (instead of K_y / K_a), for typical geosynthetic interfaces and, for side slope inclinations used in current landfill design practice.

4) The magnitude of slip along a geosynthetic interface depends on the square of the period of the base motion. Thus, the accuracy of predicted slip is heavily dependent on the accuracy of the period of the design base motion considered.

5) The normalized slip plots can be used in engineering practice by converting the earthquake ground and landfill waste motions to equivalent uniform motions according to Seed et al. (1975) procedure.

This research program continues to progress and the authors are currently investigating slip displacements induced under earthquake motions.

ACKNOWLEDGMENT

The authors gratefully acknowledge the National Science Foundation for the support of this research on seismic response of geosynthetic interfaces.

APPENDIX I   References

Newmark, N.M., (1965). "Effects of Earthquakes on Dams and Embankments," *Geotechnique*, 145(2), 139-160.

Seed, H.B., Idriss, I.M., Makdisi, F., and Banerjee, N., (1975), "Representation of Irregular Stress Time Histories by Equivalent Uniform Stress Series in Liquefaction

Analyses," *Earthquake Engineering Research Center*, EERC 75-29, Univ. California-Berkeley.

Yegian M.K., Yee Z.Y. and Harb J.N. (1995), "Response of Geosynthetics Under Earthquake Excitations," *Proc. Geosynthetics '95*, Nashville, TN.

Yegian M.K., Yee Z.Y. and Harb J.N. (1995), "Seismic Response of Geosynthetics / Soil Systems," *GeoEnvironment 2000*, ASCE Specialty Conference, New Orleans, LA

Yegian M.K., and Lahlaf A. M. (1992). "Dynamic Interface Shear Properties of Geomembranes and Geotextiles.",*J. Geotech. Engrg.*, ASCE, 118(5), 760-779.

Yegian M.K., Marciano E.A., and Ghahraman V.G., (1991). "Earthquake-Induced Permanent Deformations: Probabilistic Approach", *J. Geotech. Engrg.*, ASCE, 117(1), 35-50.

SUBJECT INDEX
Page number refers to first page of paper

AUTHOR INDEX
Page number refers to first page of paper